D0914542

WHO IS

WELLNESS

FOR?

ALSO BY FARIHA RÓISÍN

How to Cure a Ghost

Like a Bird

WHO IS WELLNESS FOR?

An Examination of Wellness Culture and Who It Leaves Behind

FARIHA RÓISÍN

HARPER WAVE
An Imprint of HarperCollins*Publishers*

This is a work of nonfiction. The events and experiences detailed herein are all true and have been faithfully rendered as I have remembered them, to the best of my ability. Some names, identities, and circumstances have been changed in order to protect the anonymity of the various individuals involved. Though conversations come from my keen recollection of them, they are not written to represent word-for-word documentation; rather, I've retold them in a way that evokes the real feeling and meaning of what was said, in keeping with the true essence of the mood and spirit of the event.

HarperCollins books may be purchased for educational, business, or sales promotional use. For information, please email the Special Markets Department at SPsales@harpercollins.com.

FIRST EDITION

Designed by Bonni Leon-Berman

Library of Congress Cataloging-in-Publication Data has been applied for.

ISBN 978-0-06-307708-9

22 23 24 25 26 LSC 10 9 8 7 6 5 4 3 2 1

This is dedicated to

Samia, my apu, you saved me.

This is for my comrades

&

to land, earth, and water

protectors & defenders.

Thank you.

We are in a space without a map . . . on shifting ground. Where old habits and old scenarios no longer apply. In Tibetan Buddhism, such a space, or gap between known worlds, is called a *bardo*. It is frightening. It is also a place for potential transformation.[1]

—JOANNA MACY

CONTENTS

PART IV: INTRODUCTION TO JUSTICE

INTRODUCTION

My life, alongside my wellness journey, has been an uncomfortable uttering at times.

I find most inner work for and about the self, especially in realms of "healing," to be inordinately excruciating. We expect life-altering transformation in a weekend's worth of work—just a skip, jump, and a head-first butt into evolution, but what they won't tell you is this game of healing is a messy business. It requires a tenacity of self, a willingness to walk into the fire, again and again. True healing means that, at some point, you're willing to die, cyclically, only to be resurrected as your truest self; it's an understanding that the journey doesn't end, you don't just one day find enlightenment, much like how you don't *just* arrive at happiness. It's not a destination as much as it is a state of being that needs consistency as well as a desire to adapt. To *become* again and again and again, this state of unraveling is the nexus point for change.

I am one of those rare masochists who loves hard work. My specific astrological placements would explain that through my Capricorn stellium, i.e., specific placements that have determined my love of climbing a never-ending mountain—right, right to the top—dedicated to the self and evolution with a reverential hunger for spiritual edification. I always played it off as a quirk, but, in recent years, as I've come to understand myself more wholly, I've begun to question what came first: Was it nature or nurture that made me so determined to get better, *be better*? Sadly, the more I unpacked that question, the fractured layers of a childhood I was so quick to forget, eject, and escape became clearer.

Some of us, maybe many more than are willing to admit, don't have pleasant memories of our childhoods to return to. The more I sought answers in my own life, the more the question became more global. If a lot of us can relate to immensely traumatic childhoods, what does that say about us as a species? As a society? Shouldn't this warrant more questioning and investigation?

When I was young, I thought I was the only kid with a bad life. I saw other children have simpler family dynamics, yet all I knew was raging catastrophe. Violence was a norm, because I had a mentally ill parent and my days were saturated by my mother's shrill screams. If I wanted freedom, it wasn't worth challenging. My mother proved time and time again that if she didn't get her way, she would punish you for it. She was domineering, the type of mother whose stare would knock you out, sweaty and gleaming with artificial regret for something you didn't even know you did. My childhood was punctuated with her physical violence—rolling pins, remote controls, she loved the bottom of shoes and sandals to hit me with. My body, our bodies—my sister's and mine—were hers to oppress and control. Yet the more she took away my power, the more I let her.

As the youngest, my role in the family was to keep it together. I was the sun, and I felt that for them. I felt that in how my energy diffused my mother's anger, or how taking a beat for someone else (usually my father) was a way I wielded a power that I understood I had as a child. The power of diplomacy. I'm naturally willful, which is a strange paradox. I was a sick child in so many ways, always bent with allergies, forever frozen, bloated, out of my body. Because there was no refuge anywhere, I believed that I had to adapt to my shitty life, so every year I tried to accept it, accept the turmoil, the suicidal ideation my mother's presence left me in. The way her groping fingers left my body forever in a state of distress. I didn't know peace or reprieve. I only felt an anger I couldn't express, and the more I wanted to, the more I grew fearful of doing so, inevitably shutting down. It took me a long time to understand that I was psychologically groomed and gaslit from childhood to adulthood about what kind of life I had been living. I don't blame anyone for this, and I've come to understand many of us were raised in familial dualities.

◎ ◎ ◎

On the outside we were a normal family, while inside there were multiple times (including two really dire situations) that made me realize I could die inside my own home. I was obsessed with John Wayne Gacy the year my mother tried to burn the house down. It was 2003, Bush had just entered Iraq, and I was a politically active kid, having formed the social justice club at school with my friend Manna. We went to protests and read Malcolm X's biography.

I'm lucky that I found organizing communities young, because it gave me a resource to sublimate my pain. This was around the same time I began writing *Like a Bird,* a novel full of grief after the protagonists' respective suicide and rape. My mother, that same year, like Virginia Woolf herself, walked out of our pale blue house after an argument with my father and told us that she was going to disappear and kill herself. Nobody stopped her; I knew my life would be easier if she were dead. Like Gacy, a clown turned killer, I saw my mother as a monstrous beast who tortured her victim's lives with near constant fear . . . but on the outside, in public, she was on best behavior. She was charismatic, enthralled by laughter, beautiful—my own Dr. Jekyll and Mr. Hyde. Because I saw her hidden parts, I saw her as remorseless as a murderer, a Medusa type, sick over all the bad things that life had dealt her. The bifurcation of self (the belle versus the monster) and because she is a Libra, meant there was a performance of self in public, a place where she could be free, she could thrive on the romance of that moment. Yet at home, in private, juxtaposed against her public self, she was a tyrant. I lived every day concerned by how her moods would swing, but there was no way to know. From a young age, I learned that I had to accept this burden of discomfort, this tragedy my life had become.

I didn't grow up with socioeconomic advantages that allowed me access to a therapist or healers, and back in the 1990s in Australia there were few resources and little explanation of abuse and how one overcomes it. So, instead of getting the primary care that could've saved me, my early childhood life was so loveless that the only way to confront it was to pretend it wasn't happening. If I could have no authority of my own, I would disassociate. I guess I never questioned how much of this discomfort was a necessity of my circumstance and simply relied on frying myself out of my own life's details. My favorite film from my teens was *Eternal Sunshine of the Spotless Mind,* a story about erasing memories that haunt us and leave us incapacitated. I romanticized this pain—it was the only way to understand it. How else to metabolize a mother who can't show you love, who also finds time and ways to hurt you in the interim of a beating? Some part of you has to harden or detach. I chose the latter—I forgot for my own survival.

◎ ◎ ◎

My name means "joy" in Arabic. I learned that from my father, who, whenever anyone asked, would exaggerate the *ha* in Fari-*ha* and mutter how my name made sense, pointing out that I was such a happy child. *When she came home from her first day of preschool,* he would gallantly tell anyone who would listen, *she had a hundred new best friends.* I learned young what my duty was, what my passport was to be in this family; I had to transform into my name for everybody's sake. I learned that my birth, in some ways, had sedated my mother's hysteria. As if she was some strange tempestuous, unsolvable mystery, and I somehow the missing jigsaw piece. So, I became a subdued court jester. Always happy, always bright. I really did want to befriend everyone, and I did. I was a cute kid, too, cuter than most kids. I knew this because I was always adored for certain things: my eyelashes, my hair, the cuteness of my face. My father's favorite story of me is on a bus at three, when I turned to a bunch of white grandmas who were staring at me, looked them dead in the face, and asked, "Are you staring at me because I'm beautiful?" The story ends there, usually punctured by my dad's slight chuckle. I've heard that story so many times throughout my life, the way parents always recycle stories to an audience, but whenever he'd say it in front of me, I'd sit, dumbfounded. What happened to that young girl? Where did she disappear to?

You learn to accept the black holes in your psyche, don't you? You accept them as a glitch, a fault of your own brain, mercilessly antagonizing your own ability to recount your life. My internal negging was rooted in a desire to reflect my parents' impatience with me, so much so that even in private I mimicked them. If I were to count the number of times I've rolled my eyes at myself, it would probably give you a clear indication of how I still sometimes feel about myself. Are we so indiscriminate with other people's feelings because we are so callous with our own? Then again, it's not like I wanted to remember my younger life; I regarded my past like a curse I had outlived.

When darkness resurfaced in me from time to time, I turned to art. Listening to Radiohead, Sufjan Stevens, Jeff Buckley—I found an outlet for my emotions. I believed I was sad because, despite having many friends, I was a loner. I felt inherently, psychically, alone. I felt that nobody understood my depth of feeling, and because I had no parent to safeguard me or my emotions, I was deeply lost and abandoned. Darkness made sense to me. So much of my outward persona was the happy parts of me that in private I gave myself permission to ponder life, be abstract, and nurture my sadness, which everyone else read as a threat to *their* happiness. My life has forever been used as a filler for other people's needs—this bred a different kind of loneliness, one that zero autonomy gives you. So, I read voraciously, and it became a phenomenal mirror to my experience. Everything from *Jane Eyre* to *White Oleander* developed into a receptacle of, and to, higher learning. I saw my mother in the pages—she was both Bertha Mason and Ingrid Magnussen. When you're used to violence, nothing surprises you, but others like you become a certain kind of kin.

My first real memory of my life is my mother tying herself to a train track. I am three. I don't remember where we are exactly, but I do remember that my mother is screaming that she's going to kill herself: *aami morbte chai, aami morthe jethe chai* (*I want to die, I want to die* in Bangla), she repeats again and again like a chant. My father is begging her not to, he's telling her to remember her children. I don't know where I am, but I know that I am her child. I know my sister is nearby as well, I'm trying to find her hand, I'm trying to find someone to hold on to.

I don't really remember ages three and a half to seven. By eight my memory is clear, but that's also around the time I started engaging with my body with a hate so punishing that I would spend afternoons pinching it in disgust. I was always trying to harm myself. I enjoyed falling over—scrapes and cuts were gratifying—but I was absolutely petrified by my mother's physical touch, which generally involved her hitting me with some external instrument. Or even just her hands, which she used so precisely to get the slaps or hits that burned the skin long after.

The constant rupture of a mentally ill parent feels like a missing heartbeat, a jump in your chest. It is to live in fear at all times, to expect alarm, to presume the worst. So, what happens is you begin to hide, you make yourself shrink in the corner, far beneath the covers so she can't find you. The thing is, I ran so far, I forgot where to find me again. It took me a long time to understand why loud noises trigger me; her screams, shrieks, banging pots and pans, or even the quake of dancing limbs running, hiding, from a hit, from a wielded knife have all made my nerves raw. Instead, by her depiction, we were made out to be lazy, impossibly entitled children, even after denying all our own needs and doing nothing but accommodate hers. I had figured out a formula to keep her the least activated, but her happiness was a mirage; she would taunt you with the possibility of it, but nothing was ever worthy of her good mood. She would often castigate herself like a martyr, normalizing acts of cruelty and disgust as if children need not be shielded from violence. It's as if she loathed us.

Without a mother's armor I was quite vulnerable to the world, but mostly to her. If you've never known peace, how can you expect it? We were always trying to survive the day. So much of my early life was trying not to jump out of the window, trying not to die, but I was secretly more like Amelié at my own funeral, wanting to watch my mother cry, finally aware of what a good child I was. I wanted her to validate me. I wanted her to see me. So much of who I am comes back to exactly this.

Quarrels in the household happened often and usually resulted in a steep acceleration of my mother's fury. She escalated fast, like an overwhelming rapturous wind, and was swift in her demolishing. A distinct memory I have is in my early adolescence. I was underneath a mango tree in our backyard. It was the nicest house we'd had yet, with an unused pool and a big, luscious mango creature with roots regal like a banyan tree. The home carried a different kind of eeriness, as homes often did back then, always with a slight off-feeling like it could be heralded by ghosts at any moment. I can't remember what I was doing moments before I heard a scream from my mother, then another that belonged to my sister. My sister is seven years older than me, and she was beginning to rebel in her own ways, fighting back if my mother hit or struck her. When my mother hit me, I took it, I cowered, always. By my teens I psychically would make my skin like steel so the sharpness of her hit would reverberate like a metallic jolt. I never fought back, I learned how to cry silently, I bore my sins. But my sister, now going to university, was learning she didn't have to take it. Though it was at the cost of our collective sanity, it was the first attempt at a family coup.

Though I would blur my surroundings if I heard even the faintest shrill edge of my mother's voice, there was something different about the scream that day; it was escalating, it was penetrating. Soon thereafter, I began to hear the pounce of bodies. That's what made me crawl out of my delirium; I knew I would have to intervene. As I began walking toward the door, something sounded very bad. My sister was screaming louder than I'd ever heard her scream, and it sounded like she was screaming for her life. When I came in through the back door to the suburban Australian bungalow-style dining room, my mother had a kitchen knife in her hand, and she was running around the table trying to stab my sister. It took me a while to figure out this wasn't some kind of lewd prank, and by the time I realized that my mother was trying to hurt her, she had made eye contact with me. I could see a void in her eyes.

It all happened so fast, I don't remember if my body lurched to protect my sister or if I just felt the urgency to run for my life, but all of a sudden I was also moving around the dining table, distracting her/running from her. At a certain point, I was surprised to hear my father's voice; he rarely got involved in these fights. Both my sister and I have asked him over the years as to why; his explanation is that he thought it would de-escalate things faster. I think for most of my life, I believed that as well. These days, I wonder if things could have been prevented. It's hard not to mourn childhood safety. What if my mother had stabbed me? After that incident, I had fantasized about that as well. Of sending her to jail. I didn't sleep by myself again until I moved to New York at almost twenty. I was scared my mother might kill me in my sleep.

◎ ◎ ◎

Instead of trying to fix the unfixable with my mother, I began focusing on myself, on perfecting who I was. I found myself too selfish, too unkind, so I focused diligently on being selfless and kind. Everything could be (and was) fixable, I figured, so my primary focus was *becoming the best person I could be*. Throughout my life, perfection has been my goal. I am in constant evolution, thinking of myself as a video game player hell-bent on winning. I'm not sure what the prize is, but it doesn't matter; I've always found the pursuit of ascension a most honorable quality. But it was never enough, and I never felt satisfied on any level. I was my own panopticon, always sniffing for reasons to self-critique.

This, I have since understood, is a trauma response. By learning how to elevate myself, I had convinced myself that the real issue was me. One of the major reasons violence for a child (coupled with a lack of affection or care) is so dangerous is because on some level you begin to believe it is *your* fault, no matter what rational understanding you might have, or awareness of the complexity. I believed anything was amenable by my own acquiescence, by my own sacrifice. I had convinced myself that was the role I played in all interpersonal relationships. To have any value, I had to give more.

Words like *grooming* never entered my lexicon as a child or teenager. I didn't realize it wasn't normal to be berated or made to feel small and useless by a parent, and that this was an act of conditioning. I grew accustomed to her verbal tirades, and I tried with all my might to make her happy. For years, I assumed I must have been a bad person because my mother managed to tell me that so often. My self-esteem was dwindling, nonexistent.

The ubiquity of sexual, physical, and even mental abuse and violence in South Asian communities makes it easy to dismiss it as "just the way things are." For most of my life, I accepted this excuse. To make matters more complicated, I found myself genuinely unattractive, and was socialized to feel this correlated with the way my mother treated me, my body a constant source of critique, so I bore it just to maintain the peace. I gave more, worked harder, just to get the bare minimum. Looking at my father, I began to understand that this is an immigrant tautology: become impenetrable as praxis, work hard forever as praxis. I don't think my parents have ever felt wanted. By their families, by their (adopted or home) countries, by each other. So, I picked up on the unwantedness, and I wore it, too. The introjection of their feelings (mixed with my own) of unworthiness and insecurity runs deep, and it transformed into a lighthouse. My pursuit of perfection against this resurrected me. It gave me purpose.

My parents both survived a civil war. In 1971, Bangladesh, then East Pakistan, faced a revolution, which resulted in genocide. The Pakistani military junta launched Operation Searchlight on March 25, 1971; nine months later, by December 16, a reported three million people were dead and four hundred thousand women were raped as a genocidal war tactic. In "What's Killing America's Black Infants?" Zoë Carpenter writes, "Health experts now think that stress throughout the span of a woman's life can prompt biological changes that affect the health of her future children. Stress can disrupt immune, vascular, metabolic, and endocrine systems, and cause cells to age more quickly."[1] These biological changes no doubt impact fathers as well, and Carpenter's analysis of the epigenetic effects on Black children descended from enslaved peoples is echoed in the trauma of oppressed people the world over, including my parents. Though it happened just fifty years ago, and even though both my parents had survived it, no history book I ever read as a child mentioned my familial or cultural history. I was left oblivious to my parents' upbringing and the terrors they witnessed, never fully grasping their pain or the brutality of their early existence. I did not grow up with the understanding that my father saw dead body parts outside his bedroom window when he was sixteen; I did not know my mother's father was almost assassinated before the war. I did not comprehend the ghosts that haunted my ancestors, my parents, and thus me, until much, *much* later.

In *My Grandmother's Hands*, trauma therapist Resmaa Menakem writes, "As years and decades pass, reflexive traumatic responses can lose context. When this same strategy gets internalized and passed down over generations within a particular group, it can start to look like culture. Therapists call this *traumatic retention*."[2] Along with traumatic retention, Menakem mentions another useful term that I was also struck by: *unmetabolized trauma*. Both phrases have a visceral feeling to them. Retention reminds me of bloating, when you are puffy with water or food; the feeling of immovability, the stuckness of unmetabolized trauma. But how, I began to ask myself, can we expect the body to metabolize toxicity at this level? This would be a grueling question I would only begin to fully comprehend while writing this book.

By twenty-five, I had hit a wall. I couldn't keep going. I hated fighting for myself, for my own existence. My mind was beginning to turn against me, I could feel it growing weak, I could feel myself slipping into a steady, mercurial darkness. Something had to change. I don't know what exactly triggered me, but suddenly I was sitting in the bathtub ready to slit my wrists open. I'd been edging on suicide, playing with it, for a long time, with the idea of death, as it seemed like a better horizon.

Around the same time, I started reading about epigenetics and was beginning to understand how trauma was passed down from one generation to the next. I was sitting with the reality that I had never felt safe, and I didn't see a way out of this loneliness. It was through this feeling—of carrying the weight of white supremacy, Islamophobia, other people's projections of me and my experience, but also the immense ghost of a childhood and family trauma that I was trying to escape—that I had my first major adult breakdown. But something happened to me in that bathtub; I realized I didn't want to die and that there had to be another way. It was an acknowledgment that I had to find a better way to live. In the foreword to her book *Civil Wars,* June Jordan writes, "I resolved not to run on hatred but, instead, to use what I loved, words, for the sake of the people I loved."[3] When I realized nobody owed me anything, something began to shift. It was an acknowledgment that I had to find a way to look at myself truthfully, holistically. I realized I needed to take accountability for myself. I realized I could liberate myself from this pain by, at the very least, facing it.

When I was finally able to afford therapy in 2019, at twenty-nine years old, I was apprehensive, unsure if a therapist could offer me anything I didn't already know. I was hyper-conscious of my flaws and honest about them with myself. It's telling that what I hadn't considered was the opposite. In explaining my life to my trauma therapist, what was revealed was what I had always known but had no recourse to accept. *My childhood had traumatized me, it had devastated me,* and I was being asked to inquire why. I'd never allowed myself to feel the hurt of my early life and, yet, it festered, entering into every fabric of my porous being like mold spreading on an unventilated bathroom ceiling. The impact could be found in so many parts of my physical and emotional self, as if I was rotting from the inside out.

My mother's own abuse had always been coded, but as the years went by, I began to piece together an explanation of why she had become the person she was, a necessary step in my healing. The past is important. But, when you're non-white, your lineages are harder to grasp. With mass migration, colonization, and the emphasis of assimilation for diaspora folks as well as the disinterest in facing the insidious ramifications of globalism, neoliberalism, and settlership, it's harder to track your ancestry.

As a young Muslim South Asian growing up in the suburbs of Sydney, I felt alone and abandoned. I understood injustice, and I wanted to highlight it because I cared. In retrospect, I realize focusing on injustice was a way to work toward seeking goodness on a macro level, as, in a way, I was also fighting for myself, my own injustice. I didn't know how to talk about how much I hated myself, because I couldn't understand why. Where did all this self-loathing come from? It wasn't all mine. This much I knew. In Australia, I was surrounded by white people, white supremacy, a domineering Australian-ness that relied on a "laid-back" and "no worries" attitude that was only ever afforded to whiteness. Everyone else I knew, the children of immigrants or refugees, experienced some form of isolation, of fear: of racism, of death, of rape. And though it was a pronounced sense of dissonance, whiteness was also the litmus test to success—as it has been for many of us. But if adjacency to whiteness was through literal physicality or money, I was failing.

The closer I got to this cavern of sadness that existed inside of me, between who I said I was and who I actually was, the closer I came to the truth. I began to understand that all these years what I had been feeling as a sense of worthlessness was a variant of shame. In *The Undying: A Meditation on Modern Illness*, Anne Boyer writes, "You think, when you feel bad, that you will never long for it, but in truth you do, since it provides such clear instruction for existing."[4] I thought I had been surviving, and yet, what I was really doing was hanging by a string, loosely holding myself from collapsing. I was always on the verge, and I could feel that friction in my soul. It wasn't enough anymore. I needed to channel this struggle into something bigger than me.

But it started *with me*, by acknowledging my tiny hypocrisies, and really questioning what self-care, what wellness, was for me.

This eventually led me to ask: What was it for everyone else? If it was for someone like me, pilfered from my very own culture, then why couldn't I afford it? Was part of my anger and frustration a result of the fact that I have always found it quite difficult to be "well"? And, if so, how many of us have felt that way, experiencing life at a severe disadvantage? This, then, became a bigger, more existential question: Was I the only one who was owed a good life? When I started to understand how deeply disadvantaged anyone who wasn't white or wealthy or able-bodied was in realms of wellness, I started to think more and more beyond just my own experience.

This book is an exploration of a few things, namely my own personal experience of needing wellness, while simultaneously examining the wellness industrial complex* and its failures. It deeply concerns me that whiteness and capitalism have co-opted wellness, relegating caring for oneself as a privilege when wellness should be for all. Instead, the accoutrements, gadgets, and clothes (primarily created *by* white people *for* other white people, while completely stealing other people's culture) have sustained inequality for the masses. Do we think of the wellness of the laborer to make these products? Do we think of what it means when an American university patents turmeric when it belongs to the cultures of the global South?† Do we think of the impact greenhouse emissions have in the countries that serve the West's blatant greed by making every-

* The "wellness industrial complex" is the braiding of what we know as "wellness" and the exploitative market of capitalism.

† The Indian government successfully challenged the US patenting of turmeric in 1996, leading to the cancellation of the patent.

thing that we use and buy? Who speaks for their wellness? *How can we be individually well if we aren't well collectively?* This book contends with the current paradox that exists by interrogating wellness through this lens of critical discourse. I'm asking us to question all the things we think we are, and to go deeper into who we can be. To make a commitment toward the collective good, I'm asking all of us to participate on a global level.

I see the wellness industrial complex as a modern arm of imperialism, and that's how violent it is. The urge for white people to commodify, to steal through a decontextualization of where certain practices come from—assuming that everything can be owned—is transparent to those who are actively decolonizing. The virulence of these actions shows how international organizations such as the WTO and IMF are false regulatory bodies that further enforce inequality by protecting only the West's interests. An obsession with intellectual property, while never considering *who has the right to patent culture,* is a farce that is becoming more evident.

This book is a call to action, but it's also an educational tool for all. Starting with an inquiry into my own wellness inevitably meant that I had to expand to think of other people's, too. In a society that colludes so righteously with capitalism, we have to ask bigger questions, but we also must act. Fast. In the age of climate apocalypse, this investigation is necessary because we are running out of time. The pandemic has shown us how we are at the behest of the natural world, no matter how much we believe we are superior. The more we lie about who we are, shielding ourselves from real, beneficial societal change, the more we pause our collective evolution and run faster toward our own demise. This inevitably means the more we deny our connection to the Earth, the more she quakes, hissing to make us listen. It's time we look at who we are in the shadows, and change.

In *Braiding Sweetgrass*, Potawatomi elder Robin Wall Kimmerer writes, quoting Native scholar Greg Cajete, "We understand a thing only when we understand it with all four aspects of our being: mind, body, emotion, and spirit."[5] So, accordingly, I've organized this book into four parts to structure our investigation of what wellness is and who it's for. In the mind section, I speak to the commodification of meditation, questioning who we are serving when we divorce it from its spiritual roots. In the body section, I speak to my IBS and chronic illness as a way to better understand what our own physical ailments can teach us. With self-care, in lieu of emotion, I write about tattoos and self-harm, and finding peace with a body that was never mine. And in the justice section, instead of spirit, I challenge the status quo, positioning the degrowth movement as a means toward a collective future that safeguards everyone's wellness, especially Indigenous peoples of all nations, whose information, culture, spirituality, and lifestyles we've stolen for gross profit. If you are a white-owned company that has propagated inequality by taking what is not yours, this book is especially for you.

Our society deeply lacks morality through the suppression, annihilation, and genocide of half the world's population for the gain of the white, West, and wealthy. We have to begin to understand that the future asks more of you. This planet, and the emerging apocalypse, requires your vigilance. The Anthropocene, this epoch we currently exist in, was created by the assault and domination of whiteness. A belief in superiority, when we exist with other beings, other species, is a simplistic and ethically corrupt way of being. We have to challenge why we've accepted such gruesome and unequal norms.

Ever since I was a child, I've questioned the bromides that govern us as a people. I always found it incredibly boring that we'd much rather accept our failures than change. The lack of genuine concern that so many have for others is a trauma response. The belief that we are owed a good life, and the only way to get it is to steal, is a mythology that capitalism continues to spew. But I'm uninterested in continuing to participate in such pathetic structures that presume power is only gained by force. Thousands of years of being ruled by masculinity, by patriarchal invention, have obstructed the necessity of spirit, of intuition, of love. War-mongering can only get you so far, and apparently ecocide is the limitation. But I'm interested in paving a new way, for myself and others who understand what a privilege it is to share this Earth. How immensely beautiful that we are all here together, now. I believe in us, as humans. My own life, and how I've healed, refusing to use violence as a tactic toward others who have hurt me, has proven to me that it's possible. I am my own litmus test for evolving in real time.

I hope through this book we can understand each other a bit better, and we can see the benefit of all of us committing to healing the things that ail us, together. Wellness could be a powerful tool to liberate society, and this is my life's mission. To educate others and show that there are other ways of being.

Kimmerer writes, "Restoring land without restoring relationship is an empty exercise."[6] We have to do this holistically now, which means getting to the bottom of things, but it also means a collective effort. There is no other way. This is my attempt to look at things more clearly for myself, but I hope for you, too. We can change, and I'm asking you to move toward that, toward your own wholeness, by caring for others as well as yourself, with an equal amount of effort on both fronts. That's what I believe true wellness means.

PART I

JOURNEY TO THE MIND

The unknown itself is in our own mind as well—our mind is in its largest part totally unknown to us. Therefore, it is not only a relation to the exterior world, it is a relation to ourselves.

—ROBERTO CALASSO

The radical emphasis cannot be simply on explaining the political information or claiming the right to information. Information is not enough. You have to train minds that can deal with information.

—GAYATRI CHAKRAVORTY SPIVAK

If you take the character of any man, it really is but the aggregate of tendencies, the sum total of the inclinations of his mind; you will find that misery and happiness are equal factors in the formation of that character.

—SWAMI VIVEKANANDA

CHAPTER 1

ON THE MIND

My mother has avoided consistent therapy throughout her life. In my twenties, after gently interrogating why, it pained me to hear her describe how it felt to be dulled, to *feel less* on antidepressants, anti-hysterics, or the medication she was occasionally prescribed. As a person who feels everything, I sympathized. I understood how shocking that reality might be, to be stripped of your compass, to lose your sense of being when you are forced to neutralize a part of yourself. Even if half of that self is shadowed by psychosis, at least it's your own. The sad thing is, there was a tricky and moral conundrum we found ourselves in as a family. By honoring her wishes, we disarmed ourselves to her perpetual violence. Abuse, in so many ways, exists in the governance of the mind.

Even still, I can't help but compare my life to my mother's and acknowledge the immense privileges I've had that she didn't. It's the weird paradox of accepting complex and nuanced situations. On soft days, when I am feeling tenderness toward myself, I am able to accept everything that has happened to me in my life. I can see myself with clarity and precision. I can hold all the things I've gained and lost and honor them separately. I know I've had access to many life choices she was never afforded, things that, even if she began to be aware of as she aged, she never fully trusted. For me, with a sister who was committed to "doing the (inner) work" before I knew what that was, I was thrust into spaces of thinking about the mind, of thinking about trauma more holistically. Even before I was an adult who could choose to be cynical, I believed in the miraculous power of healing.

Throughout this book, I'll keep coming back to the concept of policing the imagination. Saidiya Hartman writes about this in *Scenes of Subjection: Terror, Slavery, and Self-Making in Nineteenth-Century America,* stating, "So much of the work of oppression is about policing the imagination."[1] One of the hardest parts of healing from childhood trauma, especially parental abuse (or any kind of abuse that requires grooming and consistency) is unlearning the impact of something that gravely affected your expectations and experience of life. This is why mental capacity, strain, and processing are all important aspects of wellness. Sometimes, despite wanting to heal, people are blocked from healing because it is largely inaccessible to them. If my mother refuses to participate in society as a person who takes accountability for her limitations (and therefore refuses to take the steps that will afford her own safety and the protection of others), how do we factor it into the discourse of wellness that suggests the playing field is the same for anyone in the pursuit of a better life?

There are so many ways that white supremacy polices your imagination, and other systems of oppression use similar tactics to keep you small. Abuse is just another disadvantage, though it's rarely taken into context in terms of how its limitations might define you, and therefore immobilize you in the process of healing. In *When the Body Says No,* Gabor Maté writes, "Excessive emotional involvement with a parent, a lack of psychological independence, an overwhelming need for love and affection, and the inability to feel or express anger have long been identified by medical observers as possible factors in the natural development of disease."[2] Emotions, he explains later, directly modulate the immune system. As someone who has been sick for much of my life, and finding myself in multiple abusive relationships throughout the years, Maté's work helped me realize the mind and body are inextricably linked, and trauma creates a pattern of a coping adaptive response in both the mind and body.

Working with my trauma therapist to "locate my rage" has been something I find extremely hard to do. Sometimes it takes years, but when I finally synthesize my anger, like everything in my life, I turn bad feelings into gold, I alchemize shit into magic. While I celebrate this quality in me, my ability to alchemize shit into magic—believing it's what's enabled me to achieve anything—I've also begun to understand that forgiveness is not always *the* answer. Sometimes it's necessary to feel anger, to validate the fire within. It's always been easy for me to fight for justice, for liberation, because once again I sublimated these feelings into something I deemed meaningful. What I didn't realize was that there was *still* a lot of unprocessed anger that I was in turn directing back toward myself. It wasn't until therapy that I realized being forever pleasant and flexible to other people's needs wasn't a skill. Through accepting that, I was able to gain a holistic perspective of who I am, one that is not marred by judgment or haunted by the ghosts of my parents, but rather an acknowledgment by a person truly honest with herself in every capacity, good or bad. My own inner voice was so demanding, so mean-spirited, I had developed an acceptance and blur of it, believing that it made me stronger, and thus better.

Most of my life, if people hated me or did something painful *to me*, I always felt like I deserved it. I've assumed most people were, or are, better than me. I've assumed they were more moral than I was because I believed I was lacking in some fundamental way and therefore deserved to be punished and taught a lesson. This has kept me in a loop of always needing to prove that I'm the good guy. Happy to learn, to concede, happy to do more. I never considered that this may be something that I was told to ensure I remain small. I never once assumed that my mother was actually recycling ideas that she had of herself, that she then needed to reiterate and project onto me so she wouldn't be alone in that experience of her presumed failings. I was the little doll she placed secrets, fears, questions, ugliness, darkness, and depravity into. I carried them for her, I wore those feelings because she cracked me open and sunk herself into my spine, crawling into me like a puppeteer's hand.

Recently I've begun to question: Where does she end and where do I begin? Who am I when I don't have her in my head, when she isn't speaking for me, or to me? In order to keep me trapped, cloistered by her side, she had to make me believe that I could never leave. I left physically over a decade ago. But I'm still here, and so is she. Thing is, I'm trying to let her go completely so that I can embody who I am. That means looking at myself and examining what made me this way and the impact it's had on my health, both mental and physical; keeping a record, keeping track, to ensure my own safety.

For more than half my life I've suffered from some ailment: asthma, eczema, chronic body pain, dysmorphia, dysphoria, acute unknown sadness, IBS, general gut issues, constant flu/weak immune system, then early stages of vaginismus. I have never felt truly well. My gravitation toward wellness was because I didn't have a choice. I knew if I didn't confront what was underneath all the unraveling signs of my mind and body, I would live in this half life. Or else I really would kill myself. Either way, I had no choice, something had to give.

◎◎◎

I was bred to take a hit. My mother's rage could erupt at any moment, so I fashioned myself against her, cowering to be absolved. For vast years of my life, whenever the hits came, I let her use my body with disposability. In *Close to the Knives: A Memoir of Disintegration*, David Wojnarowicz writes about his father's abuse so rapturously: "You almost welcome the beatings because it's a show of some kind of affection—you'll take anything when you can't get nothin."[3] Not knowing when and what could provoke her violence, I considered my unworthiness the only possible explanation. Even if death was constantly on the horizon, the looming threat of being harmed was a sort of sick safety I attached myself to. The consistency was safe, because at least she was engaging with me, at least in those moments she knew I was alive.

At a young age I realized it was uncouth to talk about my ugly home life, so I hid it. In my teens, a few incidents made me shut out (and up) even more. The first was when I was about fourteen and I gathered a group of girls at my school because I was suicidal. In the space between one class and another, I sat on a concrete hill and shared details of my torrid familial existence. When the bell rang for class, everyone got up like I had done a presentation and walked away. No one talked to me about it again. The second time I went to the school counselor, roughly around the same age, and told them about the knife incident. I needed someone to offer me help so I could take it. My father was called into the school, and when I had to recall what I had told my counselor, I bit my lip against his visible discomfort and undermined myself. As I walked to the car with him later, I was ashamed that I had felt weak enough to ask for help, and stupid enough to think that it would come. The third and last time I was nineteen, just after I had had an abortion. For a Muslim girl this was especially bad news. I felt broken and was beginning to see how the fractures of my life had led me to this moment. Hoping for sympathy, I confided in my boss at a department store I worked at. She told me, point blank, that I had killed my baby when she was trying to have one herself. The three-punch hit of trying to ask for help but being met without empathy or results debilitated me.

I come from a culture that avoids real confrontation, especially with ourselves, so I abstracted my own feelings in order to dull the warning signs in my body. I hid the depth of my internal darkness, began to blanch myself and dim my dimensions. But projecting stoicism eventually backfires. As a child, I was expected to figure things out on my own. Being born to immigrants meant I was naturally expected to be hardworking and diligent, but with a mother who had various forms of mental illness, I also couldn't have needs. So, I became ashamed of needing anything, ever, as I felt like it was admirable to be a person who wanted nothing, and accepted nothing, who was just a keeper for others, a container for their pain and desires. I saw myself as a giver, never the receiver. I saw myself as the savior, never the saved. My trauma therapist, E, has pointed out that this is another trauma response. To know your function in a relationship can save you in your youth. To know the parameters in which I existed created a natural boundary for myself. My obsession with being perfect meant that what I deem imperfect—feelings like anger, frustration, rage, pettiness, or jealousy—I hid and dismissed. Disgusted by the idea that I could feel these things, I've masked myself in a "kind person" exterior, not knowing that I could just be a person, flawed and human. That it's OK to have contradictory feelings, and that it's not a battle to have no ego, but rather having a healthy ego, where you hold yourself and others accountable with the same justice and compassion. This obsession to be perfect, to be good, to be everything my mother needed, neutered my own sense of knowing what I needed at all times. The idea that someone could just love me as I am is quite a wild concept. I presumed I had to have a function in a relationship, and the concept of my own autonomy was so detached from me that sometimes it still takes a leap to remember that I can think differently about myself, that I can ask for more. This role prescribed to me as a "caretaker" doesn't need to be a life sentence. I can choose my own destiny, in all the ways.

We couch a lot of things as personality, in the definitions of *who we are not,* realizing that those terms and conditions are subject to change if they're not working for us. But we run on narratives of ourselves, of our lives—and then apply those narratives onto others, believing they are in on it or that they know us better than we know ourselves. This is also a trauma response. It was not until I began therapy that I realized I had spent my time in this strange purgatory of my mind. A place where I clearly had some kind of autonomy and yet my own desires were imperceivable to me.

The older I get, the more I realize I have to actually face my trauma. I have to name it, to say it out loud and voice the vibrations of my body and experience. Nothing can be decoded if it lives in mystery.

◎ ◎ ◎

In a criticism of Greek doctors, Socrates once quoted a Thracian doctor's observation: "This is the reason why the cure of so many diseases is unknown to the physicians of Hellas; they are ignorant of the whole. For this is the great error of our day in the treatment of the human body, that physicians separate the mind from the body."[4] With this book I want to isolate and examine each aspect of wellness—mind, body, self-care, and justice—because we need to talk to each other, and understand ourselves, our friends, and communities of healing and be honest about the things we've experienced, what pains us, in order to understand what our lessons are in this lifetime. In Iroquois Law,*[5] there is the idea of seven generations, a code that defines a moral responsibility to the seven generations (past and future) to heal, to face, to overcome so that those same lessons are not passed down. In the realm of wellness, it's quite amazing, because the act of micro healing—of healing oneself—must also be seen in the context of macro healing, of the community. We cannot be ignorant of the whole, we must endeavor to contextualize ourselves in the larger plight of humanity. Just as the synecdoche of the mind and its workings must be taken into consideration of the larger machinations of the body.

Amita Swadhin, an incest survivor, comrade, and founder of Mirror Memoirs,[6] once mentioned in an interview that child sexual abuse survivors are the best actors. Child abuse is such a terrible thing, and when it's at the hands of the parent, it's hard not to embody that messaging at your core, living your life within the

* The Iroquois Great Law of Peace is the constitution that governs the Iroquois Confederacy, a political alliance of six (originally five) Aboriginal nations located in northeastern North America, in what is now the northeastern United States and south-central Canada. The original five nations of the Confederacy were the Mohawk, Oneida, Seneca, Cayuga, and Onondaga. The Iroquois refer to themselves as the Haudenosaunee, or "people of the long house." The Great Law of Peace established the Iroquois Confederacy as a type of indigenous, representative democracy, which is widely regarded as the oldest continuing democracy on Earth.

parameters of that narrative. Whether or not this then becomes the compass with which you navigate your entire life, it affects every corner of your being. Imagine if you were raped, molested, or touched as a child—then imagine if that happened at the hands of a parent. Imagine how that would make you feel to carry that in your body for the rest of your life. Many people have to do that every day, without ever acknowledging that pain. Being a child sexual abuse survivor is a warrior's job, and the personal costs are endless.

At twenty-nine, after some memories of my own molestation resurfaced, I decided that instead of fulfilling a lifelong dream of having a child, my first priority would have to be to look at my own story, body, and life. When these memories resurfaced, I began to understand other parts of my mind, and things began to click. Every time I see a small child, I wonder who is touching them—I fear for them. It feels sadistic, and whenever I think such things, it feels shameful. Through medicine work, especially through sitting with ayahuasca, I've gained access to an entire community of others who were sexually abused as children, and many of us share this sentiment, a fear of unconsciously replicating the violence that is so familiar that at a certain point you wonder if there's life outside of these definitions. I want to end this cycle with me.

Maté's work has filled in significant gaps for me and helped me understand correlations in mental and physical illness. "I think [her] precocious intellectual development is what happens to bright and sensitive kids when the emotional environment isn't able to hold them enough; they develop this very powerful intellect that holds them instead."[7]

Throughout my life, I have felt like I feel too much for a normal person. I can feel so many different things at once, and they're deep feelings, oceans deep. Since I was a child I've had visions, I've seen dead people, I've had access to portals that I'm still trying to make sense of. Feeling, knowing, seeing this much is an incredible burden, however. People don't believe you, and, for someone like me whose very magic was quashed by my own mother, I became a victim of other people trying to make me small, to fit me into a box.

Every time I have cried in front of my mother, she has made fun of me. The most recent time, just before I stopped talking to her, she accused me of turning my sister into a slut. In a moment of deep despair, it all came crashing down and something inside of me cracked like a lightning bolt's whip. Of the memories I had to blur because I couldn't fathom their reality. Confusing flashbacks, phantom feelings, shivers, and discomfort were a sign that something bigger in my framework was being ruptured. Perhaps because I was sober, and also because I had just come out of a serious relationship, I was able to look at the blur, and a picture was beginning to form. I figured I had nothing to lose but to jump in.

The first time I had words for this gnawing, clawing sensation was when I watched *Leaving Neverland* and James Safechuck* explained that in his twenties he had never quite understood why *he* hated himself *so much*. I'd never seen it articulated so close to my own reality. Though something mighty inside of me was breaking, within that feeling I began to feel something dislodging, loosening, letting go. It was one of the most revolutionary moments of my

* *Leaving Neverland* is a film documenting the accounts of two survivors of Michael Jackson's alleged pedophilic behavior, one of those survivors being James Safechuck.

life. Here was a handsome white man who as a child was bright, confident, and sparkly. Yet as a grown adult he was deflated, morose, depressed, and disaffected. It was such a visceral *aha* for me, as a child survivor, that it was proof to me that Michael Jackson was guilty of the allegations he was being accused of. Because as a child survivor, Safechuck was speaking *to me*. From his eyes, through his weariness, the bleating abstraction of trauma's wake, he wore his grief, like me, on his body. He helped me believe in myself. He gave me an instrument to decode myself. I could understand myself now, within this template, and what he was saying was resonating in the very core of my being. He was explaining my own confusion about my life in the same way. I knew this wasn't happening by coincidence, and these aren't feelings that arrive in your cellular structure by accident. This was his truth, and it was also mine.

In my own vast memories, there's no one to blame. Some days, when I look back, it almost feels harmless, even though, now, as an adult, it doesn't feel good. Two and a half decades later, I'm angry at how these acts created an intense dissonance in my mind, as well as a shattering disruption in my body. A body I hated so much because of guilt and shame that was placed inside of me because of how I made another person feel. All those years I had been silenced, and yet as an adult I was still finding myself around people who were constantly wanting things from me without considering my needs. I was carrying an invisible disease they couldn't see, so I was doubly invisible. There's a pattern here, how people feel entitled to me, possessed by me, but when I ask for anything, I am told I am asking for too much. This pattern unfortunately enabled more abuse. The more I existed in this refrain where I felt so lacking, believing I was nothing and everybody else had value, the further away I got from myself.

As Maté writes, "The blurring of psychological boundaries during childhood becomes a significant source of future physiological stress in the adult. There are ongoing negative effects on the body's hormonal and immune systems, since people with indistinct personal boundaries live with stress; it is a permanent part of their daily experience to be encroached upon by others."[8] It's through the work of thinkers like Maté that my vocabulary around my own experience sharpened. Once I started seeing (accepting and then tracking) these patterns, understanding who I was through a more clarifying image, I began to trust my own vision of myself. I started to see that in a lot of my relationships there was a theme, and facing that helped me gather myself from this abstraction that trauma creates. There have been so many times in my life when I misrepresented myself through an incapacity to fully reveal the layers of truth that I—now through ongoing corroboration, therapy, and extensive somatic and sexual healing—finally understand. I'm no longer showing a half-baked version of myself, believing I am unlovable if I show my fullest self—the messy self, the sad self, the angry and bitter self, the traumatized, scared, and lonely self. Understanding the general composition of childhood trauma survivors has also helped me reflect that to my family. By seeing myself more fully, I am able to give them the kind of reflection and care that I never received. My father's once brilliant memory has been dwindling over the years, and I've seen the way my sister and I have spent most of our lives being chronically ill, existing within realms of illness. In the last few years, I've tried to understand that you can't talk about wellness without talking about all of this. About being unwell, and how ubiquitous that reality is for all of us.

I believe true healing and freedom reside in acknowledging the unacknowledged. We can't fully understand the impact of something until we look at it completely, holistically, unravel it layer by layer. Not enough of us do this because, I imagine, not enough of us believe we have the time or maybe even the wherewithal. As a species, we are obsessed with pain, yet we rarely talk about it. "Intense pain is also language-destroying: as the content of one's world disintegrates, so the content of one's language disintegrates," Elaine Scarry writes in *The Body in Pain*.[9] What Scarry is referring to, I believe, is the fracturing that is caused by pain; the split that it causes in the rational self is juxtaposed against the self that feels. If you were raised like I was—*to deny pain*, to pretend it doesn't exist—a split is born. But that doesn't mean it just goes away. In *The Undying*, Anne Boyer writes about her experience with breast cancer: "Pain, indeed, is a condition that creates *excessive* appearance. Pain is a fluorescent feeling."[10] Like a highlighter on your life, pain, whether physical or mental, pulses like an unforgettable rhythm. It aches, it sizzles, it lingers like a miasmic embrace.

My mental, physical, and sexual transgression started as a young child and continued up until I last spoke to my mother. For almost thirty years, I lived in fear of domination by her, and for almost thirty years I accepted it. I rose above it. I used the tactics my father and sister used, and I decided it was my commitment to God to withstand this. I nurtured myself in private and consoled myself with dreams. When violence erupted in the household that didn't involve me, I would be thrown into the ring to calm my mother down. At around five or six years old, I began to notice that I could remedy her the fastest, so I volunteered for this action plenty of times, assuming it was another one of my household roles. "You're the only one who gets through to her," my sister would tell me, and so I began to wear this duty with pride. Often I would go and sit with her or lie with her after an outburst. Her rage was all-encompassing, so I learned how to take it, scoop it into my body so it wouldn't hurt her anymore, so she couldn't hurt *us* anymore. I became a garbage dump to my mother's pain. But after a while, she figured out a direct passage into my body, where I absorbed everything from her, and I became an energetic scapegoat. It was like I was a portal out of *Donnie Darko*. With no protection from anybody, I began to disassociate from my physicality, an earthly reminder of the violence it carried. Juxtaposed against physical abuse, mental abuse is almost more insidious because its impact is harder to trace. The violence done is almost innocuous; what's left is just an unsettling feeling, like the remnants of an invisible hand on one's throat, like the feeling of being watched for prey.

There are years of my life that I don't remember. I used to think it was my fickle memory, but I've begun to allow myself to accept that it's the impact of trauma. Much of my early life is a blur, with spikes of acute clarity. I remembered what I wanted to remember and forgot all the rest. True wellness, for me, is wanting to remember it all . . . and knowing that I have the capacity to hold and heal it.

The human mind is a complex ecosystem of living within a pattern and swirling it around and around in your head so that everything in your life supports that reality, one that was potentially sold to you by your family or society. In healing circles, we call this the impact of living in a matrix, where capitalism, patriarchy, and white supremacy—as well as your own personal traumas and ancestral lineage—dictate and govern how we perceive ourselves, and thereby our own worthiness. I know a thing or two about bad messaging and the narratives we learn as children that become the narratives we replicate for an eternity. In Buddhism it's known as *samskara,* or the mental impressions and psychological imprints. Therapist Dr. Sue Johnson explains that this is an act of conditioning that determines the way we interact with not just ourselves but others. In her explanation of "demon dialogues," the particular attachment style we can get into with a partner, she recalls a patient declaring, "It's just like my mom used to say, I am too difficult to love."[11] This is a perfect example of how we create and sustain negative narratives with ourselves and then project them onto our loved ones. You could say, I'm in the early days of paying attention to the patterns, to the loops that I fall into that are just narratives of the mind. It's important to question whether you're a truthful storyteller. Can you trust your own memory? This is why, for me, being in therapy has facilitated one of the biggest shifts in my psyche.

Disintegration has its advantages. I've been broken so many times in my life, but were these moments just trying to show the cracks that needed to be addressed? Now I'm facing everything. I'm trying to get better, every day, at hearing myself more. Listening, ardently, I take time and consideration to gently hear myself, like a lover. By doing so, I'm absolving myself of centuries worth of pain.

The mind is so powerful, and I want to attempt to understand and clarify my own to harness it as a power for healing, for enlightenment, for compassion, for justice.

CHAPTER 2

ON MEDITATION

Before my sister introduced meditation to my mother and father (they've both been practitioners of Swami Ramdev throughout the last decade), I learned how to meditate through my political self-exploration, which brought me (vaguely) to my roots. I knew meditation came from my people, at the least the kind of meditation that has been absurdly popular since the 1960s, and there was something embarrassing (yet quaint) about learning about my lands through white people's interpretation. Thankfully, it was a gateway that didn't dissuade me (back in the 1990s it was the best we had), and like Malcolm X, a hero of mine, I was trying to track my origins, I was trying to understand myself culturally, ancestrally. During this self-enunciation, in an attempt to tackle the darkness that lay underneath the seams, beyond the surface of smiles and fighting for justice, I thought about meditating. I thought about what it meant to be healed. Could it be possible? I didn't know, but I became obsessed with the idea of stilling the mind. Though with my own mind, rapid and distinct, I was terrible at doing it.

The English word "meditation" stems from *meditatum,* a Latin term meaning "to ponder." In those early days, in a desire to understand life, I tried to ponder with no attachment to thought. The pursuit of meditation seemed like wanting to be immersed in a sea of nothingness, afloat on the feeling of deep connectivity to the universe and other beings, plants, elements. And, through practice and scripture, I gathered meditation was so much more than just the act or ritual; it was a state of being, one that impacts all the ways in which we interact with other human beings.

According to a *Healthline* article entitled "12 Science-Based Benefits of Meditation,"[1] meditation "enhances self-awareness," "can generate kindness," and "promotes emotional health." I see meditation as an approach to one's psychic well-being that is attained through consistent, at times grueling, work, but this in itself is a concept largely abstracted in the West—by white folk who decontextualize everything they extract. There's something in the "looking after yourself" that wellness provides, like a shield, to declare a sense of righteous betterment. The way meditation is taught in the West is like a catch-all phrase, a buy-one-get-evolution passport that relies on nothing beyond a callous enterprise that has labeled itself something grander than it is. Capitalism will do that. Fifty ways to help meditation get you laid, the paradox is a-plenty, and it's transparent.

Truthfully, the lineage to and of meditation is far more harrowing than detailed in the brochure for the seven-day retreat you've paid for. Part of the obsession of Western confluence with what it interprets as the East's discretion is its incapacity to understand our way of life while totally pilfering it. I say "our" because I feel this in my bones, this sense of being that is deeply anti-capitalist because it relies on real enlightenment, not your Instagram kind, which is a major philosophy of the East to invest in holisticism. To pursue one's own evolution, you have to understand it's a concerted process that demands your regular participation. In order to meditate, you should know the history, and you should also think of why you've never had to think of this before.

There are various layers not taken into context when the spiritual practice of meditation is distilled for a Western audience, which is, unfortunately still, code for "white." "People talk about decolonizing knowledge, one way to decolonize it is to return to the historical roots," explains scholar and historian Ali A. Olomi.[2] It's also another way the West gets a claim on its colonies, stealing not only resources, but filling museums with looted goods, books with stolen information. These are all just methods to further (and perpetually) colonize. When I was a child, my father would always ask me, "Who declared the WTO? Who created patenting, copyright laws? Intellectual property?" He was telling me to look further at who profits from the power of claiming things as one's own, overriding entire histories and cultures in the process.

Demonization is a part of the colonial project—it's the only way to ensure that we never, ever resist. It's very suspect to me that within the realm of wellness, language is also positioned to emphasize a divide between the former and "science," as if they're mutually exclusive, when in many cultures the two were braided into each other. The Muslim world invested in science (which is the backbone of Western science), yet as much investment was put into astrology, health, and the protection of the body. Ibn-Sina, a Muslim polymath, was known for his five-volume encyclopedia, *The Canon of Medicine*, "which includes in its section on 'Preservation of Health,' information on eight types of massage techniques and their effects and techniques." Ibn-Sina systematized the medical information gathered from dispersed ancient Greek sources and updated them with his own observations and practice.[3] Non-Western cultures categorically invested in personal health because they understood that an individual who was well was indicative of a holistically well society. There was a larger desire to protect people, and therefore healing was accessible and for all, whether it be via medicine people, healers, shamans, or witches. I remember talking to a friend from the Ojibwe nation about what's rarely ever talked about when we talk about colonization is the loss of our health.

In a *Research for All* paper entitled "Decolonization of Knowledge, Epistemicide, Participatory Research and Higher Education," Budd L. Hall, a Canadian professor, and Rajesh Tandon, an Indian practitioner of participatory research and development, explain, "What is generally understood as knowledge in the universities or our world represents a very small proportion of the global treasury of knowledge. University knowledge systems in nearly every part of the world are derivations of the Western canon, the knowledge system created some 500 to 550 years ago in Europe by white male scientists."[4] The West didn't position itself as an authority by accident, it did so by destroying the playing field to ensure nobody else could compete. They wanted to present themselves as the only purveyors of esteemed knowledge, and what this required was murdering half of the world. Which they did.

This process of dispossession of other knowledge is what Boaventura de Sousa Santos, a Portuguese sociologist, has called *epistemicide*, or the destruction of knowledge. Colonial thought relied upon destroying or extracting knowledge from the cultures it colonized, and according to Hall and Tandon, the four epistemicides can be characterized as:

1. the conquest of Al-Andalus, and the expulsion of Muslims and Jews from Europe;
2. the conquest of the Indigenous Peoples of the Americas started by the Spanish, continued by the French and the English and still underway today;

3. the creation of the slave trade that resulted in millions being killed in Africa and at sea, and more being totally dehumanized by enslavement in the Americas; and
4. the killing of millions of Indo-European women, mostly through burning at the stake as witches because their knowledge practices were not controlled by men.[5]

The most egregious part about this appropriation is that within the context of appropriation for Western benefit, an aspect of colonialism that is less frequently explored is the sadistic, protracted long game of colonization. How sick do you have to be to want to murder, steal, pillage, and rape for your own benefit . . . for an eternity? Why does no one talk about the savagery of the colonizer, only the inherent existence of that in the colonized? Meditation is a perfect entry point into looking at how something becomes taken, diluted, and then decontextualized and sold back to rich white people at a steep price.

According to archaeologists' discovery of meditation, it was first encountered in the Indus Valley, via wall art dating from approximately 5000 to 3500 BCE, with "descriptions of meditation techniques found in Indian scriptures dating back around 3,000 years ago."[6] Written evidence of meditation was first seen in the Vedas* around 1500 BCE, and is seen as a spiritual exercise, as well as a religious practice, in both Hinduism and Buddhism. "Six hundred

* "The Vedas are a large body of religious texts. There are four Vedas: the Rigveda, the Yajurveda, the Samaveda, and the Atharvaveda. Each Veda has four subdivisions–the Samhitas (mantras and benedictions), the Aranyakas (text on rituals, ceremonies, sacrifices and symbolic-sacrifices), the Brahmanas (commentaries on rituals, ceremonies and sacrifices), and the Upanishads (texts discussing meditation, philosophy, and spiritual knowledge)." "Vedas." Wikipedia, accessed October 2021, https://en.wikipedia.org/wiki/Vedas.

years before the birth of Christ, at the time when Buddha lived, the people of India must have had a wonderful education. Extremely free-minded they must have been. Great masses followed him. Kings gave up their thrones; queens gave up their thrones. People were able to appreciate and embrace his teaching—so revolutionary, so different from what they had been taught by the priests through the ages. Their minds must have been unusually free and broad," wrote Swami Vivekanada, a Bengali Hindu monk, writer, and educator, because "he [Buddha] invited everyone to enter into that state of freedom which he called Nirvana."[7]

The history of meditation and its surfacing in India is an important and often overlooked component to the historical tradition of meditation itself. Swami Vivekananda brings up two important facets of the history of meditation. First, this practice—and honestly any form of spiritual enlightenment—was governed by caste in India, and therefore generally a requirement for those who identified as Brahmins, i.e., priests and spiritual teachers, and thus a practice that excluded lower-caste, i.e., Dalit or Adivasi folks (who are seen as unable to have the same access to God) across the subcontinent. But, with the advent of Buddhism, through the teachings of Gautama Buddha, a rarefied idea of nirvana—a state acquired through focus, detail, and precision of the mind through meditative practice—was introduced to all. In fact, the Buddha was known to be openly anti-caste. According to renowned Dalit scholar, economist, and reformer, Bhimrao Ambedkar, untouchability (Dalits were previously referred to as "untouchables") came into Indian society around 400 CE, due to a struggle between Buddhism and Brahmanism.

In a paper titled "Buddhism and Caste" by Professor David Dale Holmes, "According to the Buddha, there are no distinguishing characteristics of genus and species among men, unlike in the case of grasses, trees, worms, moths, fishes, beasts, birds, etc. The Buddha goes on to show that the apparent divisions between men are not due to basic biological factors but are 'conventional classifications' (samanna). The distinctions made in respect of the differences in skin colour (vanna), hair form (kesa), the shape of the head (sìsa) or the shape of the nose (násа), etc., are not absolute categories. One is almost reminded of the statement of the scientists that 'the concept of race is unanimously regarded by anthropologists as a classificatory device. . . .' So when Buddhism asks us to treat all men, irrespective of race or caste, as our fathers, mothers, brothers and sisters or as one family, there seems to be a deeper truth in this statement than that of a mere ethical recommendation."[8] Understanding this context is important in understanding the revolutionary introduction of enlightenment for any or all—and perhaps how meditation was passed—and reached its global spread—through the Silk Road around the fifth or sixth century BCE.

If we don't understand its origins, we deny the truth. We then also have a lackluster understanding of how practices, as well as humans, evolve. It's important to note how long Indian civilization has been thinking about, and tracking, the mind. The cultural and spiritual genesis of meditation grew out of inquiry about the state of consciousness and what lay behind the guise of what we do and do not know—but also who are and how we came to be. In "Psychiatry, Colonialism and Indian Civilization: A Historical Appraisal,"[9] academic Shridhar Sharma explains the potential philosophy behind this deep inquiry: "The ancient Indian emphasized the theory of unity of body and soul and also explained how to deal with the health and mental health problems in a psychosomatic way." What an indelible word, *psychosomatic*. In Indian culture, there was an understanding of the importance of one's wholeness, that the unification of oneself—mind, body, and spirit—was a necessity to understand how to be human. Yet, despite these tectonic discoveries, the spiritual advancement of Indian thought in a modern context has been co-opted to serve Western minds and egos. It's absurd that capitalism—the antithesis of how Buddha came to be—should commodify his work, and it should be questioned, *Who is that for?* If it's in the interest of the market, it can never be for the people. In this late stage of capitalism, that much is true.

When we examine colonization, we rarely think of the insidiousness of its impact on racial capitalism, and how it disarmed multiple civilizations by neutering them of their knowledge—and then later by profiting off the exact same cultural knowledge that was destroyed. Omer Aziz writes in the *Los Angeles Book Review,* "Before the British occupation, India was not a poor backwater, but a culturally and economically prosperous civilization that had existed for millennia. India was home to the oldest university in the world, had originated our numerical system, had produced countless thinkers, philosophers, poets, and scientists."[10] My father, Dr. Sami Hasan, in his upcoming book, gathers research from esteemed Indian scholar Dadabhai Naoroji, the founder of the "drain theory," which suggests what Aziz eloquently synthesizes. Before colonization, India was a strong, leading civilization.

"Resource drainage due to colonial pillage created a huge economic and psychological dent among the victims." According to Naoroji, Britain drained forty million pounds per colonized year, creating "extreme poverty, destitution, and degradation."[11] He also calculated that the drain possibly would have amounted to £723,900,000 in the next thirty years, concluding, "So constant and accumulating a drain, even in England, would soon impoverish her. How severe then must be its effects on India when the wage of a laborer is from two pence to three pence a day."[12] Naoroji lamented, bringing up an integral point: *Who gets to have a good life and be well?* Lord Curzon, before becoming the viceroy, asserted in the British Parliament that India "was the pivot of our Empire. If this Empire lost any other part of its dominion we could survive, but if we lost India, the sun of our Empire would be set."

Consider all this and turn to *Forbes,* which declared that in 2017, 365.55 million Indians lived in multidimensional poverty. This is more than the entire population of the United States. When we unpack wellness, it's important to name not only the commodification, repurposing, and repackaging of Indigenous cultural knowledge for a white Western audience, but also to contextualize it, because through these two actions we can actually hold the entire truth. If every colonizing nation were to pay back its reparations, the entire "Western world" would collapse. Think about that. Think about what it means to be stepped on, taken from to keep alive the countries that have to dehumanize you in order to continue to rationalize exploiting you. For what else is white supremacy?

To go back to epistemicide—which involved destroying education in the holistic sense, meaning: self-education, self-awareness, and self-sovereignty—these knowledge systems were demonized, denigrated, and thwarted because they were seen as threatening to the status quo. Let's not forget that the ethnic cleansing of millions around the world required the backing of the Christian church. This was another way cultural and religious information (which was often one and the same thing) was further quashed, literally demonized, satanized. What we don't talk about when we talk about colonization is Christianity. The evidence of epistemicide is proof that the entire intention behind Western colonial power was to *create* "third world countries" by depleting them of their own natural resources that they should be allowed to feed their own people with—whether figuratively or spiritually. Yet even that—the ability to maintain the well-being of people—was stolen to ensure generations and generations of "unwell people"—a.k.a the developing, or third, world. Meanwhile, these nation-states, reeling from the pressure of orchestrated wars, genocide, and large-scale resource grabs (India included), accepted the narrative that the West propagated, only to be forced into paying off debt that these Western countries issued in the first place. If you think this is past tense, think about how often we interrogate everything the West has gained through plundering and stealing, still, to this day?

Though many of us know these things to be true, we rarely think, or consider, that Western information is propaganda sold to us as "rational, empirical knowledge," codified then backed by universities and institutions, i.e., more white men, who categorically create supremacy among themselves. What else, pray tell, is canon? "Most historians, who depend mainly on western writers, forget that the basic purpose of any colonial administration is to exploit the local resources and to improve the economic status of the imperialist power. In no country was development a priority of the imperialist power,"[13] Sharma adds. Once, a good friend's dad looked me dead in the face and said, "Colonialism had its benefits." When I asked him what they were, he told me about the railroads built in India. He was a graduate of Harvard and Yale, been a Fulbright scholar, and worked as a top litigator in a Canadian law firm.

The famous observation from Joan Didion is "We tell ourselves stories in order to live." I think white people tell the story of a mythology of their own sanctity, while they dictate colonial "discovery" by murdering half the world's population. To have a mission statement about saving, when all the impact that's been left is war, ravage, mass poverty, and disconnection from the land and self, it makes you question whether such a society values truth and honesty.

Historical archives in India (and I'm sure most colonial outposts relate) prove that the British contributed largely to diseasing Indians. "It is a matter of record that before the arrival of the British East India Company in 1700, there were no institutions to keep the insane as family supported the individual who was sick and needed help," Sharma writes. "The early establishment of mental hospitals in the Indian subcontinent reflected the needs and the demands of European patients in India during the period. Later, the development and growth of mental institutions reflected both the interest and neglect by the colonialists who ruled India for over 200 years."[14] We have to think of this also in the construction of the family unit, which in India was multigenerational. But among other things, colonizers also destroyed that, wanting insulated societies so we rely on governments to protect us, when all they show us is how little they care. If, as my friend's dad claims, there were supposed benefits of colonialism, then, categorically, why does the history of the colonized unilaterally claim (and prove) that colonialism largely never helped? That it was, in fact, inhumane and deeply barbaric, its impact leaving a torrid and ugly remnant in our legacy as a species. He's a lawyer who follows English law—he should know that the burden of proof for murder always rests on the prosecution.

This comes back to meditation, because the mindfulness movement in the West is a very wealthy one. *The Guardian* reported a few years ago, "Meditation apps monetize mindfulness; Headspace's revenue is estimated at $50m a year and the company is valued at $250m. These enterprises cater to Big Business, with which it has had a long history. Silicon Valley has a ball producing profitable, hi-tech, marketable mindfulness apps as 'brain hacks' for which there is no evidence they are helpful."[15] Capitalism, again, proves to be more valuable than human life.

In the same *Guardian* piece, McMindfulness is explained: "Instead of letting go of the ego, McMindfulness promotes self-aggrandizement; its therapeutic function is to comfort, numb, adjust and accommodate the self within a neoliberal, corporatized, militarized, individualistic society based on private gain."[16] A few years ago, Ron Purser and David Loy's article "Beyond McMindfulness" caused a defensive stir as they argued that a "stripped down, secular technique" of mindfulness originating in Buddhism not just failed to awaken people and organizations from "the unwholesome motivations that Buddhism highlights—greed, ill will and delusion," but it reinforces these behaviors. Capitalism makes you complacent, we all know that, and mindfulness relinquishes meditation of its affiliations to dharma.[17] Buddhism lite; you get all the surface without the substance.

Epistemicide happens not only with the destruction of knowledge, but also the privileging of certain schools of thought, by disguising them as precedent or, again, but in a different way, as "the way things are." Nobody questions precedent (except, I guess, lawyers), so it's important to pay attention in order to see the larger map of how everything is interconnected. As much as it's been blurred, we owe it to our society to tell the truth. This is what it really comes down to, *truth*. "Indian psychology realized the value of concentration and looked upon it as the means for the perception of the truth."[18]

If you meditate, I want to ask you why you do so. Throughout this book, whoever you are, I want you to really step in and somatically sit with your body. If you feel uncomfortable, welcome, that's important. To be uncomfortable means there's room for growth. Bhikkhu Bodhi, an outspoken Western Buddhist monk, has warned us: "absent a sharp social critique, Buddhist practices could easily be used to justify and stabilize the status quo, becoming a reinforcement of consumer capitalism."[19] It's important to question why meditation itself is being commodified and sold to people when it should be free and accessible to all.

Part of its commodification is that meditation came to the mainstream West through celebrities. "Transcendental Meditation (TM), which a 1975 *Time* story called a 'drugless high,' became popular among no less than The Beatles. As one way to cope with the strangeness of their global fame, they turned to TM, eventually going to India to study. Mia Farrow also went to India to meditate with the Fab Four after her divorce from Frank Sinatra, to study with Maharishi, whom *Time* called 'the groovy guru.'"[20] The cavalier way the West has portrayed India and thus Indians is a telling reminder of how the colonial project also determines stereotypes, which then become narratives we accept about ourselves and what role we then play within society. If there are determined roles for who is the aggressor and who is the aggress*ed*, those lines are enforced by the constant, nagging domination of white supremacy. Determined, much like race itself, as a way to classify superiority versus inferiority, it requires an acceptance that the rule book is written by people who fundamentally believe white people are *just* better.

◎ ◎ ◎

Decolonizing the mind, in India and the greater South Asia, is reckoning with a deep and painful history. As I began writing this book, mass farmers' strikes in India—against three farm acts that were considered "anti-farmer" by many—began and ended. This is just after the Citizenship Amendment Act (CAA), which allows a direct path to Indian citizenship for refugees coming from Afghanistan, Bangladesh, and Pakistan, was passed—but only for those who are Hindu, Sikh, Jain, Zoroastrian, Buddhist, or Christian. Muslims from those countries are not allowed in. "The measure is 'the first legal articulation that India is, you might say, a homeland for Hindus,'" Pratap Bhanu Mehta, one of India's most prominent political scientists, told the *Washington Post*.[21]

The dangers of any kind of nationalism, especially one based on faith, endanger anyone who doesn't comply. "The first and foremost thing that must be recognized," Ambedkar wrote in *Annihilation of Caste* in 1936, "is that Hindu society is a myth. The name Hindu is itself a foreign name. It was given by the Mohammedans to the natives [who lived east of the river Indus] for the purpose of distinguishing themselves."[22] Arundhati Roy, in *The End of Imagination*, adds, "Even today, to properly secure its idea of a Hindu Rashtra, the BJP has to persuade a majority of the Dalit population to embrace a religion that stigmatizes and humiliates them."[23] There are many anti-Indigenous, casteist, and Islamophobic policies in India that don't protect those communities of their sovereignty and right to be, which is an important part of the ways in which colonization has also impacted the relationships and diversity of India itself. Another way to disarm a population is to make it turn on itself.

The emergence of a right-wing faction of Hindu nationalism in India is a postcolonial pawn by the British to divide India along religious lines, increasing a forced tension that propelled extremism on both sides. Like any group that relies on supremacy, a brotherhood is born to attract the masses. India, a vast and magical nation of cultures and faiths of varying degrees, is now being refashioned into a colonial mythology relying on division. In *No Master Territories,* Trinh T. Minh-ha writes: "Intimidation is part of the omni-revised strategies of power exerted in the production and legitimation of knowledge."[24] As a child, my father would remind me, "They conquered and settled in India, in Bengal to be exact, because they knew exactly where and how to destroy us. They knew we would be too powerful and hard to control if they didn't." This is why context is important, everything is connected.

Capitalism, which is really racial capitalism, is the opposite of abundant thought, because it relies on the division of labor, a role everyone must know and fall into. To destabilize civilizations and bleed them of their spiritual and cultural resources so they never properly come out from under the very large and ugly shoe of colonialism is to slowly, knowingly commit genocide. That is the legacy of colonization, *legalized murder.* Rabindranath Tagore, India's first Nobel laureate, once spoke to this: "Her social ideals are already showing signs of defeat at the hands of politics, and her modern tendency seems to incline towards political gambling in which the players stake their souls to win their game. I can see her motto, taken from science, 'Survival of the Fittest,' writ large at the entrance of her present-day history—the motto whose meaning is, 'Help yourself, and never heed what it costs to others'; the motto of the blind man, who only believes in what he can touch, because he cannot see. But those who can see, know that men are so closely knit, that when you strike others the blow comes back to yourself."[25] The next part is perhaps one of my favorite things ever written. "The moral law, which is the greatest discovery of man, is the discovery of this wonderful truth, that man becomes all the truer, the more he realises himself in others. This truth has not only a subjective value, but is manifested in every department of our life. And nations, who sedulously cultivate moral blindness as the cult of patriotism, will end their existence in a sudden and violent death."[26]

◎ ◎ ◎

I'm going to ask you again: *If you meditate, why do you do it?* In 2013, Purser and Loy wrote, "Unfortunately, a more ethical and socially responsible view of mindfulness is now seen by many practitioners as a tangential concern, or as an unnecessary politicizing of one's personal journey of self-transformation."[27] A lot has changed in the almost decade since that piece was published—the world is in a different, shifting place, and we are in a pivotal time in our species' history. We are facing an environmental apocalypse as we barely survive a global pandemic; the death toll, as of this reading, is over six and a half million worldwide. In America, homelessness is up, as is poverty, exhibiting the cycle of a depleting, failed society. Capitalism is not saving us, and it's necessary now for us to look toward the future by looking back and paying homage to these traditions that conceptualized the possibilities of collective health. It's imperative that we understand that the sign of a healthy society is how we engage and look after one another. "Take the I out of illness," Malcolm X once said, "add W and E, and you have wellness." Mindfulness is not a one-sum game, because who are you being mindful of, and who and what are you being mindful *for*, if you don't know what you're practicing is rooted in?

Our position has to shift in order to start thinking of wellness as something larger, more robust than just a movement for the rich and wealthy. We have to shift toward a more united care, one that is inherently without greed. Wellness is a revolution—of communal support, of liberation. But this requires a few things. In the context of mindfulness, of meditation, it's first knowing the genesis of the very concept. It was always about healing oneself so that we could be of better service to society. But we have lost our sense of service to each other. We have forgotten what it means to owe each other. In *Sacred Economics*, Charles Eisenstein reflects on the economy of ecosystems: "In nature, headlong growth and all-out competition are features of immature ecosystems, followed by complex interdependency, symbiosis, cooperation, and the cycling of resources." It's important to begin to re-envision how something like mindfulness, in every action, could help create a better world. But it's also determining: Who deserves rest and relaxation? If we say those who work the hardest jobs—such as laborers, teachers, or civil servants—that would make sense, but that's not the reality. It's important to understand what's creating the gap between who is well and who is not. Then, it's important to change that.

Dr. Jon Kabat-Zinn, who founded the Center for Mindfulness at the University of Massachusetts Medical School in 1979, was instrumental in bringing mindfulness practice—without any religious overtones—to the public attention and scientific communities worldwide, and part of the continued legacy of meditation's distillation. A lot of this goes back to language; e.g., God should be seen as a neutral term, not tied to any creed or religion or nation. Disconnection from the land, from the elements, is a metaphor for our own disconnection from ourselves and the divine. To say something is good, better, less contradictory when it's devoid of faith is extremely reductive. The divine is and can be many things. But decontextualizing a faith-based practice from the faith itself is a dangerous alternative.

Indian mathematician Srinivasa Ramanujan, a devout Hindu, was said to credit his substantial mathematical capacities to divinity, and it was reported that his own mathematical knowledge was revealed to him by Namagiri Thayar, another iteration of the goddess Lakshmi. "An equation for me has no meaning unless it expresses a thought of God." We discredit the legacy of this knowledge by diluting the truth of its context and creation; by abstracting these kernels of wisdom, we are inevitably losing the entirety of its meaning.

"These contemplative practices were invented for monastics who had renounced possessions, social position, wealth, family, comfort, and work," writes David McMahan, a professor of religious studies at Franklin & Marshall College. Meditation was "a way of being in the world that is ultimately aimed at exiting the world, rather than a means to a happier, more fulfilling life within it." Meditation was an acceptance of death.

Reports of disturbing experiences during meditation appear in a number of early Buddhist writings. In the Theravada tradition, from which S. N. Goenka's system of Vipassana* derives, meditators are said to experience "corruptions of insight" that, from the vantage of modern clinical psychology, resemble psychosomatic ailments, including manic bliss states, gastrointestinal issues, and visual hallucinations. Monks in the Zen tradition have been documented to encounter "diabolical phenomena," which is characterized by involuntary movements and frightening mental imagery. Chinese and Japanese Zen masters are said to succumb to a "meditation sickness" in which the afflicted become disoriented and have trouble

* Vipassana, translating to "super seeing," is an ancient Buddhist meditation and mindfulness technique that focuses on regulating emotions through sensations of the body.

regulating their body temperatures and energy levels. Buddhist monastics in Tibet may develop "wind illness," the symptoms of which include confusion and agitation; according to a twelfth-century Buddhist medical treatise, the disorder is caused by the "three poisons of attachment, hatred, and closed-mindedness."[28] To me, these symptoms are a physical manifestation of people trying to understand and comprehend their mortality. It's a practice of the East, to strive for humility, to be genuflect toward God. But these days God itself is such a faraway and complicated figurehead that the notion itself has become isolating. And yet, contextually, it's nonsensical to present meditation without the context of God and the divine—and without the fear of death and the tenacity of the soul for absolution. Now, with capitalism as a false God, our motivations for self-betterment are attached to outward and public validation. This is a beast wellness has bred, but it's inevitably an unsustainable model. Without a sense of service, the sense of accepting death—which, ideally, is what the presence of having a relationship to the divine provides you with—spiritual practices like meditation are shrouded in misinformation. Death is a constant, and faith relies on the remembrance of the impermanence of life. The action of meditation is to reach a state of consciousness that's outside your connection to mortality. It's about the beyond.

I keep returning to something my friend Amber Khan, a Tarotist and astrologer, once said to me: that when the British first sent an envoy to scope the lands of India, the soldier in charge saw sadhus levitating and walking on water. I'm slowly beginning to see that the depletion of the mythology of our worlds, the worlds my people come from, was meant to rob us of our inbuilt magic, our power, our connection to ourselves. What capitalism, patriarchy, and white supremacy want you to believe is that you don't have a responsibility to think deeply about your existence. It encourages listlessness, severing from instinct, and exaggerates a devotion to self that is based on accumulation of material gain with no focus on the spiritual. It prioritizes self-optimization at the expense of community wellness.

◎◎◎

Michael Yellow Bird, a citizen of the Three Affiliated Tribes, which include the Mandan, Hidatsa, and Arikara Nations, is a dean and professor at the University of Manitoba, where he studies and teaches Indigenous mindfulness and neurodecolonization, which he describes as "combining mindfulness approaches with traditional and contemporary secular and sacred contemplative practices to replace negative patterns of thought, emotion and behavior with healthy, productive ones."[29] In a paper entitled "Decolonizing the Mind: Using Mindfulness Research and Traditional Indigenous Ceremonies to Delete the Neural Networks of Colonialism," Yellow Bird writes: "The effectiveness of mindfulness has groups as diverse as Fortune 500 companies, the U.S. Marines, Police, and Adult and juvenile prisons offering formal mindfulness instruction to members of their organizations."[30] In the same paper, he shows graphs of the mind in sacred prayer, which is a similar state to meditation, and the results are evident. When we gather all the bits we deleted and strayed from, the parts of us that our ancestors were killed for and were forced to forget, the parts we purposely discarded as an act of survival—with thoughtful consideration, mindfulness and meditation become a reckoning of those parts, too. It's not about *not* using these practices, it's about not profiting off of them for individual gain.

Language matters; science, wellness, and spirituality once were deeply rooted to and connected with one another and used to benefit larger society from the dark perils of the human experience. There is time for us to return to ourselves, and we need to understand that this is a collective fight. Meditation for the masses. In his 1907 speech "The Spirit of Japan," Tagore spoke with prescience, about moving away from capital and greed, to return to a sense of the ideals around community and a sense of greater service: "For when this conflagration consumes itself and dies down, leaving its memorial in ashes, the eternal light will again shine in the East—the East which has been the birth-place of the morning sun of man's history. And I offer, as did my ancestor rishis, my salutation to that sunrise of the East, which is destined once again to illumine the whole world."[31]

For me, as a child of the East and West, as someone who has heavily been involved in decolonization for almost two decades, it's important to track these histories. The ones that have largely never been told. By doing this I'm not only beginning to understand the cultural lineage of my people, and how that has impacted my life through epigenetics and the phenomena of processing trauma, but I'm also understanding my parents and their failures. The mind is a strange and tenuous place, but is it really as mysterious as we make it out to be?

As a child, maybe due to my Venus Aquarius in the eighth house, I was obsessed with deeper inquiry. I always wanted to know more; now I'm following the trail. I think it's a collective responsibility to do this now, to look deeper into our minds and at the very least acknowledge what they are telling us so we don't have to keep gaslighting ourselves or the people around us. In some ways, meditation is about that, too. Accountability is accepting death, because it's accepting that you're fallible. Meditation is allowing you to come to this state with zero judgment yet full responsibility. You can't diminish such a sacred offering and expect it to capitulate to capitalism.

CHAPTER 3

ON INTUITION AND UNSEEN THINGS

I was seventeen, maybe almost-just eighteen, short-haired and earnest, when I started working at David Jones, an Australian department store in Sydney, in the men's suits department. In between the lull of the customers, I would chitchat with folks in other departments, especially the people in accessories, the section right next to me. My social life was limited, so work often felt like socializing.

One of my favorite coworkers was a forty-eight-year-old Italian-Australian breast cancer survivor and mother of three. Her name was Carla, and she spoke with passion, angst, and a soft fragility that was very moving to me. What I first noticed about Carla was that she was changeable, one could say moody. She had leopard-printed blond highlights and spoke with a dramatic twang. I was changeable, too, struggling to feel ownership over my body, not knowing what my worth was when I wasn't being desired. Maybe Carla saw that in me, saw my body's discomfort. One of her daughters was a punk lesbian, and I was at the age where I was longing for a mother, especially one who would teach me things. In our fleeting moments of reprieve, during the long hours on our feet, we would lean against the cash register inside our panopticon, surrounded by gaudy Guess wallets, and talk about life. One day we were cleaning out the corners of her kiosk, bent over the cascading variation of navy- and turd-colored accessories for men, when she shared that she had once suffered from a severe eating disorder. I listened, finding relief that another woman had also hated her body as much as I did mine.

For so long, I didn't know it was a shared plight. Though I went to a girls' public high school, I assigned power to everyone else around me, never once assuming that anyone *could* hate themselves as much as I hated me. So to hear Carla share her belief that her latent yet domineering self-hatred—all the ways she felt betrayed by her body because it wasn't thinner, or more white—that she wasn't blond-haired and blue-eyed, or delicate like another kind of white girl (I wonder if she even felt white), was the major reason that all these years later she had gotten breast cancer. She believed that the anger she had toward herself had metastasized into cancer in her body. That day she cried, softly, into a tissue. Though she was speaking to me, I felt she was really speaking to herself, maybe assuring herself of the truth in front of an audience—so she couldn't take it back. Later I wondered if she'd ever admit it to me again.

This moment of Carla's vulnerability would become a reference point for me throughout the years. I would return to it like an oracle, like an alarm. It became a moment in time so significant, so unparalleled, in the formation of self that it would change the very way that I understood time, space, and the consequences of the mind on the body. I was also ripe for this perspective, as I had already started seeing patterns within myself that were beginning to make me question my own self-destructive behavior.

I started cutting myself in high school. Never anything too sinister, but I enjoyed the way a blade, a safety pin, or sewing pin (which was sharper), or even the side of a steel ruler's edge felt erotic against my skin. A release is a release—and oh how the pores on my body pulsed as I lined it with an incineration. I liked the way a bandage felt on a wrist afterward, a deluded but necessary act of care. Only the other day I was sharing this tidbit with my therapist, adding that this is why I have long believed I now have, and enjoy, many tattoos—it's sublimating pain into resilience, grief into art. My first tattoo was when I was nineteen, a big year for me. I began getting tattoos to absolve my body of its sins, to carve myself a body I could be OK with. A body that was mine alone, determined by the markings that brought me closer to myself. This, my therapist has pointed out, was another way to show up for myself, to show up for my own pain, by nursing my wounds and tending to them after the act.

At the time that Carla told me about her theory, my sister had introduced me to a book called *The Hidden Messages in Water* by Masaru Emoto*. I was beginning to comprehend the immense impact of energy and the very essence of it. My own body was a testament to self-hate, and so perhaps that's why something clicked about how our own minds affect our very tangible physical health. I resonated with Emoto's almost utopic conception: "When you have become the embodiment of gratitude, think about how pure the water that

* Emoto's research has been widely criticized for his use of unconventional methodology. Detractors have suggested that because his results are not replicable using the standard scientific method, they are fraudulent, while supporters have argued that the scientific method cannot account for the unknowable, spiritual dimension of research.

fills your body will be. When this happens, you yourself will be a beautiful, shining crystal of light."[1] It would be years until I would gain further clarity about how my negative thought process was internally destroying me, though it was a formative introduction.

In his 2005 review of *The Hidden Messages in Water* for the *New York Times*,[2] Dwight Garner writes, "Every so often a bestseller comes along that, as Michael Korda put it in his book *Making the List*, makes us 'question our sanity, or at least that of the American reading public.'" I was very conscious of the way people viewed things of a spiritual nature, even as a teenager. My sister beckoned me with strange self-help books that my father, an academic, would snort at, dismissing her entirely. Instead of championing her radicalism, my parents mocked my sister's quirks. Privately I supported her, but in front of them I participated in bullying her. I understood, in the rationale of the scientific world, her spiritual, psychic, and emotional meridians were baseless. My father's obsession with information, cataloging it like a Rolodex, was his primary motivation in life. To this day, my dad's ability to reference academic text dumbfounds me, but back then, there was something he wasn't considering: the world of magic, the world of spirit. And that his daughter was ahead of her time or, rather, tapping into something deeper. My sister trusted in what she was being drawn to. She's the first person I ever witnessed to truly listen and stand up for herself.

The domineering patriarchal conception insists science is only what we know, without interacting with the possibility that there are also a lot of things we don't understand—*and a lot more we are yet to know about*. That must shock people who want to believe in their own hubris. Western science clearly has a history of being racist and exclusionary, and throughout its conception there have been many examples. Among them is John Harvey Kellogg,* an American medical doctor and eugenicist, who was also a major leader in progressive health reform, particularly in the clean living movement, a period of puritanical crusades across America.[3] Kellogg, who looked like a sanitized Colonel Sanders, was a man who "chose to focus instead on improving the so-called superior white people, the Anglo-Saxons. Teaching them the proper principles of 'biologic living,' including whom to mate with, and what and how much to eat and exercise, would ensure supremacy in this life and the next."† In her book *Fearing the Black Body: The Racial Origins of Fat Phobia,* Sabrina Strings, associate professor of sociology at UC Irvine, notes that a lot of twentieth-century science and medical journals were determined by an obsession to uphold and protect whiteness, and thereby proliferate anti-Blackness by demonizing Black women's bodies. "Interestingly," Strings writes, "the relationship between fatness, disease, and inherent racial deficiency described in the medical literature did not appear to grow out of scientific findings but rather resulted from sentiment and impression."[4] Fatphobia (and thereby the demonization of fat bodies as being "lazy" or "uncouth") is deeply interconnected with the history of white supremacy, and thus the subjugation of Black bodies in response.

* He is also the brother of Will Keith Kellogg, the founder of the Kellogg Company, as in, yeah—the cereal.

† "A clean living movement is a period of time when a surge of health-reform crusades, many with moral overtones, erupted into the popular consciousness. This resulted in individual, or group reformers such as the anti-tobacco or alcohol coalitions of the late twentieth century, to campaign to eliminate the health problem or to 'clean up' . . . The movements also coincided with episodes of xenophobia or moral panic in which various minorities were targeted as undesirable influences for medical or moral reasons."

"Clean Living Movement," Wikipedia, accessed September 25, 2021, https://en.wikipedia.org/wiki/Clean_living_movement.

But this sentiment is far more complicated and insidious. In a modern context, we know that in America, pharmaceutical industries have immense control over healthcare. According to the Harvard Public Health paper "Snapshot of the American Pharmaceutical Industry,"[5] "The United States is the worldwide leader in per capita prescription drug spending, representing between 30 and 40 percent of the worldwide market." The paper goes on to explain, "Further complicating the relationship between the pharmaceutical industry and patients is the fact that drug manufacturers in the U.S. are allowed to market their products directly to patients. Such practices add to the overall amount of national drug spending, since many of these drugs can be quite expensive." The proof is in the pudding, as they say, and the more I traced the dots together, the more I started to understand that American and global medicine is just another industry trying to maintain the status quo: whiteness, patriarchy, and capitalism. It's a market that needs to be fed, so I wanted to know—how (and why) do we keep feeding it?

◎◎◎

In writer Cheryl Wischover's essay "Can Wellness Be Scientific?" for *The Cut* Charles Mueller, a clinical associate professor of nutrition at New York University, says "It's very difficult for people like me to keep my head on when I hear about things like alkaline diets and detoxing. There's no such thing as detoxing your body, absolutely no such thing."[6]

I could easily overlook a relatively innocuous statement like this until I started to think of how the statement was simply untrue.

Undermining the cultural (e.g., Ayurveda*), ancestral (e.g., sweat lodges), political (more on this in a second), and spiritual (Ramadan, Lent, a day of penance during Yom Kippur) action and knowledge that has been passed down, and has centuries worth of cultural proof behind it, puts us in a strange predicament when it's done by someone with authority, who, like Mueller, has the backing of an institution. The more I started to think about how writing about wellness coalesces with those in power and therefore determines who gets to speak on the subject with authority, the more I realized we were in a trap of whiteness. In this instance, science is coded as "right," but any system that governs with impunity should be questioned.

Perhaps the article's question is also at fault, as it insinuates that wellness and science (again) are incompatible, or are both devoid of the other, without considering the dearth of knowledge that remains rooted in different Indigenous technologies (some might even call them "sciences") that make up for the majority of modern wellness. "What is known through intuitive (ancient) knowing and unnerves the rational (scientific) mind is often dismissed as unprovable, irrational, or simply wrong," Sebene Selassie, a Buddhist teacher and writer, documents in *You Belong*. She adds: "Power is wielded by those who master language and argument as well as access and resources."[7]

As I kept thinking about the importance of intuition and energy in traditional wellness practices, I began to see how the mischar-

* In Ayurveda there is *panchakarma* (*pancha* means "five" and *karma* means "treatment"), which outlines a process to prepare the body for detoxification and then five methods to remove toxins. It's an important element of an Ayurvedic lifestyle.

acterization and dismissal was wielded by authoritative voices that were coded in "science" and the rationale of the West, which had to be accepted through force. Thanks to wellness marketing, the mechanics of wellness have been adapted into the capitalist machine, creating a system where people are still unconscious, still investing in individualism, without believing in the possibilities of community, openness, vulnerability, and a willingness to be well with each other. The many failures of the wellness industrial complex are the result of the disinformation it relies on, by negating the historical context that it leeches from.

In her book *Women Who Run with the Wolves*, Dr. Clarissa Pinkola Estés explains the way women have lost their sense of agency through the ages. "Early in the formulation of classical psychology women's curiosity was given quite a negative connotation, whereas men with the same attribute were called investigative."[8] She uses gender as a binary, but to me it's less men versus women than it is the attributes of one gender versus another—hence patriarchal methods versus matriarchal ones.

We could look at the binaries of science and wellness like this, where one is seen as more rational and intellectual, which is coded as masculine, while wellness is seen as intuitive, irrational, and therefore castigated as feminine. This is not a coincidence and goes back to the barbaric methods of the colonial project. "When we lose touch with the instinctive psyche, we live in a semi-destroyed state and images and powers that are natural to the feminine are not allowed full development," Estés writes. "When a woman is cut away from her basic source, she is sanitized, and her instincts and natural life cycles are lost."[9] This could be used as an analogy for the Western world and how it has maintained power through colonization, and therefore suppression of globally Indigenous lifestyles. Perhaps because when we are sick—when we are unwell, depressed, when we are confused by our own power and agency over our bodies—we are easier to control.

To go back to Mueller, though not the same as "detoxing," the merits of fasting offer incredible value. Labor activist Cesar Chavez, who called fasting "probably the most powerful communicative tool that we have,"[10] described his relationship to fasting as a "sacrifice for justice."[11] For the *New York Times*, Mariana Alessandri, assistant professor of philosophy at the University of Texas, added another layer by speaking to the gendered connotations of the act of fasting itself, and who is believed and not believed in that narrative of spiritual fasting. Using as an example the French writer, philosopher, and mystic Simone Weil, whose connection to God and commitment to empathy was seen as spurious and untenable exaggerations, Alessandri writes, "In many of the subsequent accounts of her life, Weil has been labeled an anorexic,* like St. Catherine of Siena, the 14th-century nun who was also convinced that she could reach God through her empty stomach. This was not the fate of Chavez, who is today still admired for carrying out a centuries-old ascetic practice, nor of Mohandas Gandhi—nor for that matter, Jesus Christ, who in the Bible is said to have fasted for 40 days and nights in the desert."[12] It's important to highlight these stories because I believe the reason we uphold "science" and occlude "wellness" also impacts how we feminize and gender certain modalities. In her essay, Wischover cites professor Kevin Folta, chairman of the horticultural sciences department at the University of Florida, who quips, "I don't remember who said it first, but the plural of anecdote is not data."[13]

* The correct term might be *Anorexia mirabilis*, which is now seen as a form of religious asceticism that largely affects Catholic nuns and religious women.

The suggestion that the efficacy of ancient wellness modalities is purely anecdotal, without considering the vast histories—thousands and millennia longer than "Western thought"—that have brought us this information is a classic tactic of Western science. To mystify without engagement is a silencing tactic. It is also an imperial invention to esteem one source of knowledge over the other. To present an argument about something as expansive, illustrious, and profound as the world of wellness is simply a way to delegitimize one's own argument by showing an archaic diligence to speak loudly to assert power. According to Alessandri, and thousands of years of spiritual practice as backing, "Some research now supports the idea that the body does well when it gets deprived of food for certain lengths of time."[14] As Chavez reminds us, fasting has tremendous reparative value, because being temporarily without food allows the body to heal.

White men of course love making grandiose statements in journals and newspapers, postulating their fancy ideas with unbridled sarcasm, like "Emoto's book is filled with photographs of the shiny, happy crystals that form when water is shown phrases like 'You're cute' or 'Thank you,' and also of the deformed, sorry-looking crystals that result when you show it 'You fool!' or 'Satan,'"[15] from Garner's review. The apparently sarcastic slant is replicated in the *Irish Times* review (several years later) by William Reville, who writes, "There is always an audience for this kind of thing. Some people just want to believe in strange phenomena. Emoto's followers are also likely to be into New Age phenomena like chakras, out-of-body experiences and past lives."[16]

Pay attention to language. Pay attention to the way the insults are ladened. It's an important pronouncement, and it makes me wonder: Amid the conquest of patriarchal invention, colonization, and white supremacy, why do (white) men still feel the need to demoralize? To humiliate what they don't know? Using the calling cry of the nihilist, they've weaponized critique as a way to thwart all ideas outside their own purview. Never questioning what encourages them to insinuate that intellectual rigor can only be found in their own limited ideas of the world, and that nothing else requires their focus or attention. How pathetic to live like this.

◎ ◎ ◎

The word *chakra* is derived from the Sanskrit word meaning "wheel of light" or even "disk" or "vortex." The earliest (known) written record of chakras is found in the Upanishads, circa the sixth century BCE, and the concept of the chakras exists in two parallel dimensions: one is the physical body (*sthula sarira*) and the other the energetic, known as the subtle body (*sukshma sarira*). This subtle body consists of energy channels (known as *nadis*) that are connected by portals of psychic energy, which is what we call a *chakra*. In early Sanskrit literature, the psyche naturally corresponds with the physical body, and they mutually affect each other. According to Indologist David Gordon White, the synthesis of the chakra energy centers was introduced around eighth-century CE in Buddhist texts, and in Theravada Buddhism the *cakka* was seen as a wheel that encouraged temporal dispensation through the wheel of dharma. "The Buddha spoke of freedom from cycles in and of themselves, whether karmic, reincarnative, liberative, cognitive or emotional."[17]

Unsurprisingly, the chakras are also similar to the meridians in Traditional Chinese Medicine (TCM), and *qi* and *prana* are two respective words to mean the "energy of life." TCM relies on a similar concept of chakra, but the channels are called *meridians,* and similar to blocked chakras, symptoms of various illnesses are often seen as the impact of disruption, deficiencies, or imbalances of energy, or *qi.* Chakra imbalances may also be caused by stress, diet, trauma, lack of exercise, and disconnection from self or purpose. As *prana,* like *qi,* flows through the top of the head and down to the chakras, each chakra nurtures a distinct form of energy related to that specific aspect of what that energy center means. In correspondence, there are twelve main meridians in TCM. Each limb has six channels, three yin channels on the inside, and three yang channels on the outside.

There are seven chakras all together, and this is how they merge with TCM:

1. The root chakra is located at the perineum (Meeting of Yin). The spleen (yin) and the stomach (yang) are represented by the earth element and the first chakra. This chakra grounds us in the physical and is related to security and survival.
2. The sacral chakra is located between the kidneys, which is called Mingmen (Gate of Destiny), and on the front of the body just beneath the navel, called Spirit Palace Gate. The kidneys (yin) and the bladder (yang) are represented by the water element and the second chakra. It governs creativity, vitality, sexuality, hormones, kidneys, the urinary tract, and reproduction.
3. The solar plexus chakra is located in the soft space beneath the sternum called Spirit Storehouse and on the spine directly behind called Jinsuo (sinew contraction). The heart (yin) and the small intestine (yang) are represented by the fire element and the third chakra. This is the seat of our emotions. Blockages can manifest as anger, sense of victimization, poor me, and resentment.

4. The heart chakra is located directly in front of the heart (Center Altar) and on the spine at a point called Shendao (Spirit Path). The lungs (yin) and the large intestine (yang) are represented by the metal element and the fourth chakra. Blockages can manifest as heart or immune system problems, or even a lack of compassion.
5. The throat chakra is located in the lower part of the throat called Heaven's Chimney and at the base of the neck on the spine (C7) called Big Vertebra. The liver (yin) and gallbladder (yang) are represented by the wood element and the fifth chakra. The throat chakra is all about communication. Blockages manifest as insomnia, inflammatory conditions, stress, tension, migraines and/or skin problems.
6. The third eye chakra is located between the eyebrows at a place called Yintang (Hall of Inspiration) and is called Wind Palace. Third eye chakra is linked to the pineal gland and is connected to the higher self, intuition, and psychic abilities.
7. The crown chakra is located on the top of the head and slightly back at a place called Hundred Convergences (or Meetings). It is the connection to higher realms and is associated with helping congestion, sinuses, and in calming the nervous system.

It is said that when chakras are blocked, emotional and physical problems begin. The conceptualization of the chakra system is to understand ways to heal the body, mind, and spirit, and in doing so, bring the chakras into alignment and balance. In a similar way, meridians describe the overall energy distribution system of Traditional Chinese Medicine, which helps us to understand the notion of carrying, holding, or transporting qi, blood, and body fluids around the body. We are more than just our physical body. We are made up of a system of various levels and dimensions of subtle energy that are interconnected and interdependent. This knowledge is rooted in the concepts of meridians and chakras and is important to acknowledge in the future of our society and of our species. The more we think expansively about the body—and its capacity—the closer we can get to truly healing because it contextualizes our existence, again, in a macro way, which means integrating the mind.

In *The Gift of Freedom,* Mimi Thi Nguyen writes, "Ideas of gender, race, and coloniality are central to these assumed scenes of liberalism, and to the global empires that found liberalism's emergence."[18] In a nebulous way, capitalism—alongside the impacts of race, class, and colonization—has determined who gets freedom over their bodies and who doesn't. I think of the concept of Taylorism, named after the nineteenth-century management consultant Frederick Taylor, who designed a method of optimizing workflow in factories, which then influenced classroom design, introducing the idea that children's bodies needed to be optimized and controlled.[19] This creation has severely impacted various realms of civil society—for example how we prioritize "work" and thus value, and also how we demonize others for "laziness," i.e., the rhetoric used against Black people and fat people to further propagate the idea that a palatable whiteness was the apex of evolution, and everyone else's responsibility (if they don't fall neatly into whiteness) is to serve the idea of it. The more you cause a society to remain unhappy in who they are, believing that their value is dependent on their output, the more a void is created, which can easily be (momentarily) filled by leaning on products to fill a hole. If you can't afford those things, or provide a service that is deemed "worthy," the more you are forced to accept your inherent lack of value and submit to your position of servitude to the state, to capital, believing—maybe on some level—you don't even have a right to be. Bring in abuse, trauma, as most humans have experienced some level of terror. Unloving parents, abusive partners, random acts of violence. Then bring in war, displacement, genocide. On top of this, let's add in systems that are supposedly designed "for you" (governments, for example), because what is democracy if not an entity that serves the people? Why else does the United States invade other countries if not for the projection of democracy? So, what happens when that very system lies to you, subjugates you, imprisons you, steals from you? There's a two-parter here: (1) We have to begin to understand that these systems that we've been told think about *us*, care about *us*, provide services to safeguard *us*, don't actually. Because foundationally they can't. I think about this in an energetic sense—when something is built on lies, how could that house of cards not eventually collapse? And (2) That it's time to accept that we can demand more, but only when we identify the problem.

"Medicine tells us as much about the meaningful performance of healing, suffering and dying as chemical analysis tells us about the aesthetic value of pottery,"[20] Ivan Illich writes snidely (but accurately) in *Limits to Medicine*. We've put Western science and medicine on such a pedestal, which therein causes a dangerous slippery slope. "The unexamined assumptions of the scientist both determine and limit what he or she will discover," as the pioneering Czech-Canadian stress researcher Hans Selye pointed out. "Most people do not fully realize to what extent the spirit of scientific research and the lessons learned from it depend upon the personal viewpoints of the discoverers."[21] Like all systems, Western science and medicine, too, are not objective just because we say they are.

◎ ◎ ◎

"Disease is never neutral. Treatment never not ideological. Mortality never without its politics," Anne Boyer writes in *The Undying*.[22] I've kept returning to this statement again and again, especially as Carla's words about her cancer have stayed with me these many years. Sometimes people are able to articulate the exactness of an experience, and at the time when she and I were in each other's orbits, I don't think I was aware of how much my own self-hatred had governed me. "It is a sensitive matter to raise the possibility that the way people have been conditioned to live their lives may contribute to their illness," Gabor Maté writes in *When the Body Says No*.[23] In reference to the memory of Mary, a Native woman in her forties who suffered from multiple diseases, he asks: "What creates this civil war inside the body?" A question that no medical professional had asked. Nobody had wondered what this Native woman might have experienced, the sociocultural realities of her own life (e.g., how missing women are ten times more likely to be murdered than other demographics[24]) or even how the epigenetics of a brutal endless genocide can live in your body as you remain unseen and unnoticed by the government that has claimed and settled on *your* land. Isn't it important to understand a body's holistic, psychological state at the onset of the disease? It's what Socrates mentioned millennia ago: Where is holisticism in care? And why isn't it valued? "Mary had been abused as a child, abandoned and shuttled from one foster home to another," Maté writes, adding in her own articulation, "I was so scared all the time, but as a seven-year-old I had to protect my sisters. And no one protected me."[25]

When you are a survivor of childhood trauma, the entire world asks you to hide your pain. I have seen that people don't want to be burdened with your reality, especially people who have had relatively good lives. So, they ignore your truths, your dimensions. Every friendship I've lost in my adult life has shown me that in most cases, people don't want to afford others the complexity they want for themselves. I imagine all of this comes back to scarcity, a belief that there isn't enough, that we all should be in competition with others. Deflecting real matters, we are forced to turn on one another.

I am also realizing, more and more, that I'm writing for people who feel their trauma every day but desperately want to heal by looking at it. I'm writing for people who have seen darkness in their lives, too, and have felt misunderstood for the weight of what they hold and carry. There is something bigger to work toward, and we need each other to do it. Being truthful and open about my experience helps me to gain a better perspective on my mother's life and all the factors that made both my parents who they are. Over the last year, I've gained a lot of insight from my father about why he never intervened and his complicity in my mother's violence. Like Mary, I protected my sister and father because I always knew, because I was told, that I was the one who could ease her off the ledge. So, my body became the soft toy to throw into the ring. I latched on to these roles well into my adulthood where I was the caretaker, reenacting my abuse again and again. Hoping for love and feeling as if the only way I would be loved was if I sacrificed all my needs and gave everything away. Then maybe somebody would love me.

"We tame our natural physical expression from an early age. We learn to police ourselves," Selassie writes. "Through our families, communities, and society at large, we absorb messages about what is acceptable and not acceptable in terms of our existence: how to move, how to speak, how to dress, and how to belong or not belong."[26] There are so many details that go into becoming, into feeling like you belong to yourself, to your family, to society—or even to love—and so much of that comes back to how we are conditioned as children, then in life. It's interesting to me how we, in the West, have allowed the veneer of capitalism—money and all that it can buy you—to become the logical reasoning for humankind, and not the investigation of the spirit and soul. It's fascinating that this also, again, correlates back to a very gendered understanding of masculinity. The investment in nation-state, border, domination, warmongering are all masculine traits in a very rudimentary sense. This is the same way rigidity, a lack of emotionality, and rationality—as well as a hefty dollop of cynicism—are also seen as masculine qualities, and thus rewarded in a society that upholds disconnection from what is right or fair, and rewards who can strong-arm the other more. Holding the scarcity-driven belief that power is not innate but stolen.

When I remembered, or rather turned and faced the light, the haunting ghost of my childhood sexual violation at the side of my neck, breathing on my ear forever, something just clicked, like I had always known but had finally strung the words together. I think this is always the first step of healing from abuse; it requires you turning to face what is looking at you. I put my eyes on the memory and saw what I had always known but was too afraid to say. My body carried the physical signs of my mental anguish for so long. It was in the countless UTIs, the countless yeast infections, a body always filled with mucus. In my parents' home, I would wake up in my sister's bed every morning with a pocket of saliva filling my mouth. I was constantly sneezing, coughing, and blowing my nose—but only ever at home. In the outside world I could breathe.

Society doesn't support this kind of mental discovery, as there is very little support for children of sexual abuse. Yet according to a study on the impact of sexual abuse on female development,[27] on average, we showed signs of cognitive deficits, depression, dissociative symptoms, maladaptive sexual development, asymmetrical stress responses, high rates of obesity, more major illnesses and healthcare utilization, dropping out of high school, persistent post-traumatic stress disorder, self-mutilation, *Diagnostic and Statistical Manual of Mental Disorders* diagnoses, physical and sexual revictimization, premature deliveries, teen motherhood, drug and alcohol abuse, and domestic violence. In addition, "offspring born to abused mothers were at increased risk for child maltreatment and overall maldevelopment."[28] I knew my body had been trying to communicate that something was very wrong for a very long time. At twenty-nine, I was ready to face the mental work needed to heal myself.

The way we deny our connection to the Earth is the same way we deny connection to ourselves. This happens in multiple ways when you are in the margins of society—but especially if you are a woman or femme person in this world—you are taught not to trust yourself and forced to accept the rules by which everything is governed, even when those rules weren't made with us in mind. As a child I was encouraged to mistrust myself. If I was hurt, physically or mentally, I was told to get over it; even when my dad was being sweet, he was still being dismissive. The best of us do that to each other—deny our truths—and sometimes we do it to remain within the confines of the status quo, sometimes because the idea of facing something feels too big. I think back to a life before I accepted certain things about my childhood, when so much was unknown to me about myself but was right there, staring back at me. If I were to look myself in the eye, I would've had to do the impossible, blow up myself and my family. So instead, I stayed quiet, played by the rules, and made myself so small so nobody would question anything about who I was. But it was only a matter of time, I think I always knew that.

In *Caliban and the Witch*, Silvia Federici writes, "On the one hand, new cultural canons were constructed maximizing the differences between women and men and creating more feminine and more masculine prototypes. Women were inherently inferior to men—excessively emotional and lusty, unable to govern themselves—and had to be placed under male control. As with the condemnation of witchcraft, consensus on this matter cut across religious and intellectual lines. From the pulpit or the written page, humanists, Protestant reformers, counter-reformation Catholics, all cooperated in the vilification of women, constantly and obsessively."[29]

We have to begin to understand the feminization of wellness—by denouncing it through the language of being hokey or woo-woo—and insinuating that Western science is the be-all and end-all of information needed to understand who you are is a deeply dangerous template. That allotting intuitive understanding of yourself and your body is not, nor should be, your primary concern. Much like how God has been co-opted by a mediator like a priest, one's own well-being has been co-opted by the lingering miasma of medicine and science that is, in its totality, not even invested in understanding your specificity as a human. Our bodies are merely data.

Civilizations that were destroyed by the advent of epistemicide and the colonization of the world were societies eager to expand, to know more. Historian and scholar Ali A. Olomi explained about the Muslims: "They had a deep, deep appreciation for ancient knowledge. They had an understanding that the ancient world had secrets and mysteries that needed to be preserved. So Persian forms of knowledge, Indic forms of knowledge, Greek forms of knowledge had to be preserved."[30] There was an inherent appreciation of the vastness of information.

In a paper on Cartesian philosophy, Enrique Dussel, an academic and philosopher, reframes Rene Descartes's statement "I think, therefore I am" into "I conquer, therefore I am." Descartes, who introduced the concept of dualism, was a strong believer that the mind and body are two different metaphysical substances, believing that the mind can exist outside of the body, and the body cannot think.[31] The conceptualization of the mind-body problem, as dualism initiated, is an alarming reality for men like Descartes who were trying to understand the complexities of the human condition within, again, the confines of control and discovery. As opposed to acceptance and understanding. Recently I found myself working with a new massage therapist. After our first session, she explained that my QL (quadratus lumborum) was more engaged on the left side of my body. "It attaches from the top of the pelvis to the bottom of the ribs," she noted over email when I asked her for further explanation. "What I've found is that when this muscle is more engaged on the left side, the person feels like they can't present as their full self; like they have to make themselves smaller in some way to be accepted or received in certain situations." A muscle told her everything she needed (also "kneaded," *sorry*) to know. As Bessel van der Kolk writes, *the body keeps a score* of the mind's workings.

I think part of unlearning these vast and failing systems is to learn to trust ourselves and our own wisdom, but we must also challenge the status quo and unpack how we play into domination. Holding ourselves accountable and listening to our morality and our gut is key, but then we must act. If we are holding ourselves to the highest standard, then it's easier to trust that the people you allow in are also going to be *as* invested in holding themselves with as much humility as you are. If that's not the case, question your relationships, because this restructuring is the cornerstone in the creation of a care network. In envisioning a future of care, disability justice advocates are some of the best thinkers reimagining a future for all.

In *Care Work: Dreaming Disability Justice,* writer and educator Leah Lakshmi Piepzna-Samarasinha writes, "Crip emotional intelligence is understanding isolation. Deeply. We know what it's like to be really, really alone. To be forgotten about, in that way where people just don't remember you've ever been out, at meetings and parties, in the social life of the world. How being isolated, being shunned, being cut off from the social world of community is terrifying because you know that it can literally kill you. And that being alone also does not always have to be killing; it can also be an oasis of calm, quiet, low simulation, and rest."[32] We have the ability to accept the inevitable and change. To open, expand, and shift to listen to ourselves. In order to hold ourselves and each other better.

I think about that a lot, about Carla, about my mother, about myself, and the grief that comes with realizing that all those years you talked to yourself with disdain were because you were told to mistrust who you were. For me, that meant accepting violence, but sadly, that's a common occurrence. Many of us are unlearning from our painful pasts. Many of us were forced to be silent, to be small. For everyone who has been made to feel like a burden by our families, the marginalized, the abused, the broken but whole—it's time to move toward a new pattern of being. With each other.

Estés writes a utopian vision for who I want to be: "The medial woman stands between the worlds of consensual reality and the mystical unconscious and mediates between them. The medial woman is the transmitter and receiver between two or more values or ideas. She is the one who brings new ideas to life, exchanges old ideas for innovative ones, translates between the world of the rational and the world of the imaginal. She 'hears' things, 'knows' things, and 'senses' what should come next."[33]

In a lot of ways, this unique time, where we are on this earth, is calling us to return to ourselves, in ways we have never been called to before. This means returning to a way of being that is intrinsic, connected, and holistic. Not just with ourselves, but understanding that as a society, we can't be fragmented and disconnected from the needs and realities of each other. "Care is feminized and invisibilized labor,"[34] writes Piepzna-Samarasinha, but imagine if it was a reality for all. We need to get to a place where we can learn how to trust ourselves again. We need to understand that how we demonize intuition, women's knowledge, and wisdom is deeply entrenched in the colonial project, thereby directly impacting how we engage with wellness, care, and the world of healing. It's all connected. It's time now we finally faced it.

CHAPTER 4

ON THE FRUSTRATING PITFALLS OF HEALING

As I said in the beginning of the book, healing is fucking excruciating. The ups and downs of the mental turmoil—all the sick spaces your memory unveils—then the tragedy of finally having to face this trapped grief that lingers in balmy childhood passages that are long gone. It's an interior excavation that will deplete and exhaust you. Oftentimes I have felt incapable of continuing on this path, one where I have complete transparency over myself at all times, which means facing those dark corners when they emerge. While I trust and harness my own intuition so that I can be cognizant of the signs, of the intuition that we have been dissuaded from, the most difficult part of this journey, I've found, is the loneliness of self-exploration, of mooring the dark unbridled waters. When I started really committing to this path, it meant an acceptance of everything that was required, basically a complete soul activation to liberate myself from all the systems of oppression that bind me, that bind us, as a society. And how they all intersect in real time, in the now, affecting our livelihood in the current moment.

Piepzna-Samarasinha tells us that "survivorhood is a part of disability justice," meaning survivorship must be understood in its complexity, as a disability, as something that thwarts and impedes your life, giving you an inordinate number of hurdles. Survivors of extreme trauma are not often seen for the invisible disabilities our bodies can carry. I've noticed—especially with folks who have had brutal lives—that there's often a part of them that wants to be dictated by that pain. Even if they don't want to identify with it, their whole personality becomes a reaction to what they can't articulate fully. Then, inversely, for folks like me, people who hid from their reality for so long, the pain lingered boldly like a bleeding ink spot wherever I went, but I got good at looking away from it. Memory is complicated—it's both reliable and unreliable with such resoluteness, with such searing clarity—but trauma itself, when you get to the bottom of it, doesn't lie. It's like a bright, pulsing, probing orb. In one way or another, it's calling out to you, and one way or another, it will find you because it's so desperate to be heard. Through the radical act of listening (finally) to myself, I've begun to uncover all the things I was hiding from myself, but I've also started to understand how trauma is passed down until someone in the family decides to change the course of history by deciding to confront it.

When we talk about trauma, how often do we think of the mental work it takes to actually process such things? So many of us forget, refuse to acknowledge, or take it out on others because it's hard work that oftentimes requires returning to sites of trauma. I have been obsessed with the word *abreaction* ever since I learned of its conception; to me, in my layperson's terminology, it's about reliving trauma to release it. For me, that has meant going back and reminding my child self, baby Fa, that I'm OK. That I'm safe. It has meant using EMDR therapy* rewriting neural pathways, which I've also committed to with the use of sacred plant medicines to go to the memory, to see it, let it pass through me, to feel the energy clear. This is some of the mental work that is required when you are healing.

I keep saying language matters, but what you do with that language also matters. In a climate of the wellness industrial complex, there is an obsession with the right language. At times, looking around, it feels as though more people are invested in sounding aware than actually doing any of the work required to get the results some of us blatantly claim we're working toward. It was during my last sit with the sacred medicine Grandmother ayahuasca† that I realized we are all just patterns, narratives stitched together and presented to the world as "who we are," when the very idea to arrive somewhere permanently, into a version of yourself you just accept, is anti-evolutionary. In a conversation with Indian writer Arundhati Roy, American writer Naomi Klein said, "Too often it's been 'how can we change without changing?'" Shouldn't we want to shift, to change, to be better? Don't we need more people who are committed to being their best selves at all times? Isn't that what "doing the work" is ultimately? What I'm talking about here requires mental commitment.

* EMDR (eye movement desensitization and reprocessing) is a form of psychotherapy used to heal trauma and negative emotions caused by distressing memories whereby the patient is asked to recall those memories while simultaneously following the finger movements of the therapist with only their eyes. This is believed to help remap the areas of the brain that associate certain memories and thoughts with harm.

† Ayahuasca is a psychoactive sacred plant medicine found in the basin of the Amazon. It's taken ceremonially as a means to heal, with the help of Shamanic practices. Evidence of ayahuasca use has been dated back to one thousand years ago.

When I moved to New York, I was absolutely shocked and consumed by the capitalism of the city. Something that has always disconcerted me about Americans is their lack of compromise. It's something that as a non-American I have found quite confronting. American culture is predicated on a mythology of its greatness, of its (white) supremacy, encouraging individualism, divide, distrust, and an extreme solipsistic obsession with self, which is something that in a globalized context has become a part of the reach and perpetuation of capitalism. In *The End of Imagination,* Roy reminds us about the abuses of the American empire: "We know that while, legally and constitutionally, speech may be free, the space in which that freedom can be exercised has been snatched from us and auctioned to the highest bidder."[1]

I had my political awakening in high school, as I mentioned, but my adult awakening happened during the Ferguson riots in 2014. The murder of young boys Trayvon Martin (seventeen), Tamir Rice (twelve), and Michael Brown (eighteen) felt devastating to hold. Even though it wasn't a new phenomenon—the murder of Black and brown youth at the hands of those who are supposed to protect them (in this instance, the police and the neighborhood watch, both entities that supposedly exist to serve and protect)—in my adult life, it was the first experience I had of understanding that if anti-Blackness was to end, white supremacy would have to be confronted. Be it the neighborhood watch, be it outside a bodega, or in a car with your wife and baby daughter in the back seat—in order for this carnage to be stopped, we would have to fight for justice and face the system head-on. I've repeatedly thought of how Philando Castille's daughter will heal from witnessing her father's murder at the hands of the police. What does she do to block out the sound of the gunshots? Of her mother declaring, after her father has been shot five times, "Please don't tell me he's dead." Don't we owe both of them a healthy life after betraying them both so badly? Don't we owe more to the children whose lives were stolen from them? This isn't fate. It's state-backed violence. The mental work in confronting racism (in the world and in yourself) is a mental labor I doubt (sadly) most white people really delve into. But it's that kind of working of the mind that we need now. This is what it means to fight for justice; it's mental work first that is necessary if we want to make true, impactful change.

As someone who has worked toward justice for most of my life through organizing, through self work, through my professional work, I've learned that anti-Blackness is always present and pervasive in the context of white supremacy. This country does not have a healthy track record. It was built on stolen, and thus *free*, labor. If you look at the formation of this nation and don't feel deeply upset by it, challenge that. Is it that you have a hard time humanizing others? If you've accepted the country's past sins as historical absolutes, challenge that. Because this country doesn't look like democracy, or fairness, or justness to many. In actuality, it looks like exploitation and greed that's been snatched and auctioned to the highest bidder. What happens to a society that can't face itself? It suffers, and it remains unwell.

"The American Way of Life is simply not sustainable. Because it doesn't acknowledge that there is a world beyond America," Roy reminds us. The analogy of trauma and America is a perfect explanation of what happened during the Trump presidency. The rise of the extreme fascistic right wing (a global movement) is occurring because white people feel threatened. En masse, they, perhaps without fully realizing or understanding why, are losing what they have (stolen) to immigrants, Black people, Muslims, Jews, and all the rest of us who are vilified. But what they don't (can't or won't) accept is that what they had was never theirs to begin with. We can only be suppressed for so long. As the saying goes, you tried to bury us, but you planted seeds. The United States, with its history of war or attacking, Korea, Guatemala, Cuba, Laos, Vietnam, Cambodia, Grenada, Libya, El Salvador, Nicaragua, Panama, Iraq, Somalia, Sudan, Yugoslavia, and Afghanistan—to name a few—needs to begin the reconciliation it takes to heal.[2] That's another part about doing the work—you really have to do it, you can't just skip over the actualizing parts. It says a lot about a society that complies as it's being lied to in the face. I think of this Alice Walker quote: "People who think nonviolence is easy don't realize that it's a spiritual discipline that requires a great deal of strength, growth and purging of the self that one can overcome almost any obstacle for the good of all without being concerned about one's own welfare." In the face of such unending inequality, it's remarkable that things haven't totally collapsed. Much like wellness, it's important for us to finally acknowledge that white supremacy is a disease that must be uprooted, but by doing so it needs to be declared and diagnosed.

Since the age of about four, my body has been revolting in one way or another—yet the medical world very rarely gave me true understanding of the language of my body. It took decades before I understood that leaky gut syndrome is one of those very real conditions that medical experts routinely dismiss. I always felt that there was something lost in translation about my sickness, creating further confusion because it was being abstracted by Western medicine. I've never felt understood by the healthcare system—my needs were always too obscure, almost like I was speaking in strange, mysterious tongues.

Because my nourishment was depleted—my mother too busy with her illness; my father too busy trying to raise two kids and navigate his wife's temerity—I accepted that I was flawed. Naturally, as I watched the world around me not pay attention to my needs, I began to dismiss them as well. I didn't think I had a body; all I felt was that I was a giant blob of confusion. This was around the time I started to grow disillusioned with my gender as well. I felt unappealing, and I was putting on weight. I now know there was nothing wrong with that, but back then I was bullied or teased by my family for being "the fattest one," which in South Asian families (at least now) is code for "undesirable." I didn't feel like a girl, I felt like an alien. Desire was outside of me. Around me, I was surrounded by white people and the occasional East Asian who also felt completely out of my league. I didn't have nice accessories from Morning Glory, the cool Asian stationery store—I had shitty cheap things from Kmart. With no frame of reference on how to be, I became obsessed with being Japanese. Karen Watanabe was a foreign exchange student who was my temporary best friend. I, too, wanted to be glamorous, I wanted to be liked. The irony is that East Asians are often incredibly racist to South Asians, so again, I was confused—othered. Against this backdrop, I became more and more confusing to myself. What space could I occupy? Who could I emulate? The more this happened, the uglier I became to myself. This created a cycle of abstraction. Over time I couldn't recognize who I was anymore because I had no mirrors, so the easiest thing to accept was that I wasn't worth living, that my very life was purposeless, so I had to find some source of purpose, some source of life.

◎◎◎

In "*The Crane Wife*," CJ Hauser writes, "I had arrived in my thirties believing that to need things from others made you weak. I think this is true for lots of people but I think it is especially true for women. When men desire things they are 'passionate.' When they feel they have not received something they need they are 'deprived,' or even 'emasculated,' and given permission for all sorts of behavior. But when a woman needs she is *needy*. She is meant to contain within her own self everything necessary to be happy. That I wanted someone to articulate that they loved me, that they *saw* me, was a personal failing and I tried to overcome it."[3] I used to think it was romantic that I wanted to give everything, that this made me a good partner and thus more deserving of love. I wanted to be an easy, chill partner—someone to settle down with. The opposite of my mother. In fact, caustic women, women with immense needs, would often trigger latent anger in me. Demanding women, blunt, mean women I found upsetting and incorrigible. I knew myself to be callous, crude if I wanted to be—but I controlled those parts of myself. I saw the incapacity to be kind as a weakness. All of this is the working of the mind. Stopping patterns, tracking narratives about the self or others, and rewiring yourself to believe in your worthiness—these are mental acts that require determination of spirit.

I once read that intuition requires understanding that when something is emotional it's a feeling, as opposed to a *knowing*. That was a major turning point in my participation in the journey of my mind, of recognizing that I had more power and will than I had first determined by listening to my own intuition. When we teach children they are too emotional, we also teach them that they are deceptive to themselves. It's the same thing as gaslighting, forcing someone to distrust their own beliefs and knowing in favor of someone else's. This has happened with every romantic partner I've had—I'm gaslit to not trust myself and my own feelings and intuition. A lot of my life has consisted of people telling me I'm wrong to feel, think, or believe the way I do.

How do you form a backbone when you've never had one? How do you create the foundation in your spine to support your interior organs so you don't collapse every time you hit a bump in the road? That's been something I've been thinking about recently, how the very foundation of my being has been rotting since birth. How my family's legacy, and their brokenness from a civil war and partition completely divorced them from their sense of purpose. I've never really thought my parents liked themselves—both of them are insecure and suffer from variants of mental illness. Yet both of them have felt obliged to dictate the paths of my sister and me. My sister was the first one to hear the call of the spirits, and whatever she did, she stirred something awake in me as well. The big gaps in between where I have no memory are the moments of time in my life where I forced myself to forget how to be me and took on the role of being somebody else. I would say this happened from the first moment of abuse, until about the age of twenty-nine. Up to that point I lived in the shadows of myself. It's as if the hands of my conditioning had pulled me apart, going deep into the fabric of my skin, puppeteering me into life. So much of my existence has been for others that returning to myself felt too uncomfortable. Turning the car around takes precision, a precision that I lacked. Yet I also knew if I didn't, I'd die on the path ahead of me. So, at a certain point I had to start with the mind. I had to start healing this place first.

◎ ◎ ◎

My mother showed me that the world was mean. She could turn light to dark in a matter of mere hysteric moments. Her stare still haunts me. Her judgmental gape, eyes crudely narrow. It is not an exaggeration to say that the moments of tenderness she and I have shared in my thirty-two years can be counted on one hand, while there are so many moments of trauma, moments that pulse like open wounds. But the closer I look at them, the less they hurt. It's a funny thing, healing—the wildest realization is that it works.

These days I also remember the moments of laughter. How her dimples round her face, how her eyes grow small like a cat as she laughs and laughs and laughs. Oh God, what I'd do to make her laugh. How every ache would stop as she gullied her cheeks, fluffy like clouds, and how beautiful she was when she felt loved. I did it to save her, I wanted to give her my life. I let her win, I let her hurt me and touch me inappropriately. I always felt like I was being lusted after by her. My body was never her daughter's body, it was always just hers, and thus an extension of a body she never had herself. One of the last times I saw her, I went downstairs to eat breakfast. I was wearing an oversized T-shirt an old friend had given me. I had worn this shirt so many times and had learned to always cover myself underneath it so as not to be too alluring. It was an uncomfortable reality, being sexually preyed upon by a parent. But I was gaining more agency in myself, and maybe this is what had started to infuriate my mother. I now lived so far from her that her control was barely felt. I had too much freedom, and she had to show me who had more power. That morning, before going downstairs, I had debated putting on underwear, but I was also beginning to understand that my body was in fact my own. So, I went downstairs and entered the kitchen. Everything happened so fast. One moment I walked in and the next moment my mother had lifted my shirt to reveal my entire naked body. She stared at it, then me, and spat out, "Loja lage nah?" which translates to "Aren't you ashamed?" This is not just physical (and sexual) humiliation, it's mental as well. It took me a long time to understand that.

I wasn't sure what I should be ashamed about, so it was just easier to get small, so small so she couldn't see me. This is where the brain comes in. We all carry our PTSD, trauma, in such intricate and idiosyncratic ways. Through time, I've begun to piece things together—because I was sexually exploited, I felt unsafe and turned against my own body. I knew that I felt detached, disembodied, but I never thought about the effects of trauma on my mind, or that maybe my mind was even *more* affected, and my body was just the evidence. It's as if I felt my real self was too threatening, so I hid, because who would accept me for the person I truly knew myself to be? That disgusting person I knew myself to be.

The cycle of keeping myself small and battered became the norm. In a society where survivors, let alone child survivors, are not protected, it shouldn't come as a surprise that a lot of us have a hard time accepting parts of ourselves that were taken against our own will, and that many of us live in those dark rooms, those dark memories, incapable of knowing how to move forward. It is a hard reality to swallow, so it's easier to forget. But then you are expected to live your life, never speaking out loud about all the ways in which you have been compromised, so, in moments where that anguish does come out, you're seen as crazy because you can't hide it. Then *you* begin to pathologize yourself as crazy, because clearly you're messy and embarrassing and everyone can see it and so you enter yet another loop on the hamster wheel of shame. You let others dictate how you feel about yourself until one day you realize you're the only one who can stand up for what you know is your truth—you have to declare what has been done to you. You have to forgive yourself. This is the arduousness of healing the mind.

I could never have fathomed, when I started this specific journey eight years ago, that I'd uncover a plethora of secrets and energetics trapped in my body, but once I made a commitment to myself through therapy, all the rest began to make sense. I've noticed that the more I like myself, the more I gain peace for what I've seen and felt. I forgive myself and I forgive my mother, my family, for all of it. This is part of the work to heal—it means ultimate liberation from all the things that give you definition, that feed the trauma, and letting go. But all of this has to come from yourself—it has to come from a sense of will. Sometimes that will mean understanding that this feeling of clarity might shift and hurt may return. Nothing is stagnant, and everything is a cycle; another curve on the spiral. To accept the rise and fall of healing takes immense courage. It takes immense endurance.

By my mid- to late twenties, I realized people might act like they cared, but it was important to examine the actions behind their words. There was, again, a pattern. No coincidence, the moment my life finally felt like it was shifting in the right direction, my personal relationships threatened to destroy me. Through time, I've noticed that people like to project things onto me, and because I'm porous, I take it on. It was as if I was attracting people to treat me the way my mother always had. Cajoling, castigating, always at the whim of others. It's no surprise that the people who knew me most intimately also knew my trauma, and they then knew how to hurt me with it.

Shine *less* was what my mother told me my whole life. I know this was her own trauma, too, of always being too bright and having to learn how to fold herself into a tiny little pocket square to not be in the way. My mother is beautiful, one of the most beautiful women I have ever known. She was always told that she looked like the Bengali actress Sharmila Tagore, big adoring eyes, a sweet smile. My mother was an impeccable learner—the best cook, gardener, clothes maker, entertainer, and the shiny thing at a party, burning so miraculously. But behind closed doors she was austere and mean, overwrought with criticality, brimming with a hardness that she took with her everywhere. My life was hijacked by my mother—my happiness, brightness had to be given to her to hold up. In *Care Work*, Piepzna-Samarasinha writes, "I am very fast—in comprehension, in knowing what you're feeling, in knowing something is wrong. I have superpowers: I know what you're feeling before you feel it. I've been praised for it all my life. . . . And I have internalized that it is my job and my worth, the thing I am skilled at doing, the thing that was my value when I was not seen as worthy of protection."[4] Choosing myself over my mother has been a slow, unraveling lesson that I'm still currently processing. It often felt difficult to accept that I had to have such a boundary that she could never infiltrate my mind or compassion again. I moved all the way from Sydney, Australia, to New York City as a young person because I needed to believe she'd never find me again and all contact would be initiated by me, because a good day could turn bad in mere seconds when I was in conversation with her. It's as if I was always fighting for my life.

Most people will fail you with their incapacity to understand how childhood or familial abuse cripples you internally. I didn't think I had agency over my life because I was never allowed to have it. I felt as if my life was happening to me, not that I had the capacity to shift and change it. Naturally driven, I had the cognizance to escape a shattering home when I was barely just a young adult, but the real shit—the life force, sovereignty of self shit—I lacked. I felt completely docile, too frozen in fear to self-actualize. Instead of acknowledging that part about me, I pretended for much of my life that I was fine and that because I had survived my life, I was stronger for it. The more I drifted in this alternative version of myself, the more something—I felt, and knew—was deeply, eerily wrong.

In the past I've seemed to be attracted to people I could hand my power over to. I enjoyed being criticized, I liked being told I could be better. I liked putting people on imaginary pedestals, making myself believe I needed to serve them and gain their acceptance. I wanted both platonic and love relationships to have this kind of nature, because for me, even platonic relationships bordered on romance. I didn't know what a boundary was, or how one could help or aid me.

Eventually I realized, no matter how hard I worked, that this would not make up for the giant void that I felt in my life. I realized I had to slow down, and that I had to listen to my needs and prioritize myself. I didn't even know what that meant, but I knew I had to find a way to make my life work for *me*. That's a huge thing to utter as a survivor. Nobody tells you to do this when you come from familial abuse, otherwise how would the abuse metabolize? Everyone partakes in it, and so things are far more complicated than they seem. Not a lot of people can sit with the sadness of being betrayed by a parent, and the loneliness it feels to never want to engage with them ever again. I feel sad for abandoning my mother, and I feel like a bad person, despite knowing the decisions I took to safeguard myself from her are also the reasons I'm still alive. But, for years, I told myself I was selfish because I got to escape. The fact that I get this life, this freedom, while my sister still cares for my mother is painful. As a Bangladeshi, filial piety keeps us from straying, but what it does more than anything is it obstruct the self and create a dangerous alliance with codependency.

Finding camaraderie with my sister and my father has helped me to understand that my mother is a result of so much agony, and so, in many ways, we all mourn who my mother was in her lightness. I saw her in all her incarnations. What a sacred transaction to have with another human being. She's just a hurt person, and she has every right to be. When I remember that, she's no longer my mother. I can forgive her again and again. Or, at least, I can understand a universe where that forgiveness exists.

◎ ◎ ◎

When I started safeguarding my own mental health, many of those in my periphery failed to respond with support. As I began to create boundaries, I was often met with resentment. But the more I examined the company of people I kept, the more I began to imagine what a world looked like where I was in equal loving relationships. It's been beautiful to understand that I am allowed to feel, yet every day I still have to remind myself how to be soft. I'm also beginning to realize that I can ask for that from others, as well.

Healing is an everyday practice that requires personal resolve and diligence. The mind is such a deep and cavernous space, but our liberation from our struggle, from pain, relies on reworking, reweaving, and reimagining what it looks like to feel safe and at ease with oneself. If I wasn't on this journey, I wouldn't believe it was possible. It's weary and dark navigating and forming new neural pathways as you relinquish yourself from the bind of trauma and victimization. I'm beginning to see and understand that I can have the life that I want. I just have to take myself there slowly, gently, patiently. With all the love I never received yet had inside of me for all these years. As Maté writes, "When we have been prevented from learning how to say no our bodies may end up saying it for us."[5] My body said no a long time ago—I'm finally learning how to catch up and tell her, *I'm here.* I'm listening. Doing this work starts with training the mind. That's when something miraculous appears. For me, it's been sourcing self-respect where there formerly was none. The self-respect that's arrived from understanding everything I've been through has brought me to this great place of self-affirmation.

PART II

JOURNEY TO THE BODY

The body will always tell the story of our woundedness in a language so direct and simple that it can be too much to bear witness to. As protection, the mind pulls away and keeps itself isolated from the body. The most radical project we could ever engage in during our lives is the project of embodiment. This is the most radical act because there is no liberation without the union of mind and body.

—LAMA ROD OWENS

CHAPTER 5

INTRODUCTION TO THE BODY

The body is an inherently political thing.

I think of bodies most in terms of labor, as in which body (correlating with what labor) is deemed more valuable in society? And, of course, within a global context, what kind of body (and labor) is more valuable, *period?* There are hidden hierarchies that we seem to want to deny. So much of the modern interpretation of being human is cast against the concept of value. When you live in a colonial world, where the impacts are ripe and pervasive, the legacy lingers on every surface, around every corner, lurking. My father lived in a country for over a decade where a significant proportion of the labor force is brought in from Bangladesh, his own home country. He understands that though he came from the same place as these men, his body and class are deemed more valuable—due to the access to education, adjacency to white, privileged spaces, and status conferred through books and intellectualism.

As a society, a collective reckoning is needed about the impact of colonization on the body, so we must begin to understand the body as an inherently political thing. It's important to look at the disparities between how a disabled person and an able-bodied person are treated in the world; how class creates perimeters to health and access to basic human needs; how caste, race, and anti-Blackness are propagated and wielded in order to control and divide society only to ensure one type of body is better than another.

It's important to understand that race was concocted by whiteness out of fear, out of a need for domination. That then has a generational impact on the mind, body, self-care, and justice. In the United States, the division is so palpable because the statistics are clear: The livelihood and life expectancy of Black and Native Americans compared to white Americans is vastly unequal, and unfair. In *Love and Rage: The Path of Liberation through Anger,* Lama Rod Owens writes, "And the thing about white colonist fear and rage is that I have nothing to do with it but my body still becomes a receptacle for this unmetabolized woundedness. At the end of the day I find myself hauling not just my trauma but also the trauma of whiteness."[1] We know now about the intergenerational influence of epigenetics; so what happens to your body when you are healing lineages of exploitation, abuse, murder just because of the body you occupy? "Trauma," writes Owens, "disrupts the body's equilibrium."

In the context of the pandemic, this became clearer than it ever had been before, highlighting the depth of class disparity and racism. Abolition was called for, which was a major turning point in the lexicon of American liberation. Representation has always been important, but shrouding the truth in propaganda is also the American way. "The a priori association of Blackness with guilt and criminality comforts white America by enabling people to believe that Black Americans are deserving of their condition and that the livelihoods of whites are in no way bound up with Black immiseration."[2] What is clarified is that within the disposability of bodies—of trans bodies, Native women's bodies, Black bodies, Palestinian bodies,

Dalit bodies, Muslim bodies, disabled bodies, fat bodies, queer bodies—there is this willingness to disconnect from one's own humanity in order to dehumanize another by taking their life.

Disposability is a big theme in capitalism. It plays into how wealthy people navigated the pandemic, by either having access to better healthcare or being able to escape to safety, while the poorer (and statistically non-white) people in the cities died. In part 8 of *Under the Blacklight*, Dr. Kimberlé Crenshaw's podcast, Emery Wright, an organizer, explains, "We are living in a public health system that is not designed to ensure public health."[3] This reminds me of what Anand Giridharadas writes in *Winners Take All: The Elite Charade of Changing the World*: "American scientists make the most important discoveries in medicine and genetics and publish more biomedical research than those of any country—but the average American's health remains worse and slower-improving than that of peers in other rich countries, and in certain years life expectancy actually declines."[4]

Is it a lack of care—this complicity? How can our incapacity to think outside of ourselves be diagnosed? How can we rectify this damage? In *Close to the Knives*, one of my favorite books of all time, David Wojnarowicz explains it perfectly: "Most people tend to accept, at least outwardly, this system of the moral code and thus feel quite safe from any terrible event or problem such as homelessness or AIDS or nonexistent medical care or rampant crime or hunger or unemployment or racism or sexism simply because they go to sleep every night in a house or apartment or dormitory whose clean rooms or smooth walls or regular structures of repeated daily routines provide them with a feeling of safety that never gets intruded upon by the events outside."[5] This inequality and inconsideration has seeped into the entire reality of living in the United States. Through silence, the status quo remains; through intimidation, power is asserted, wielded, and eventually dominates.

֍֍֍

As mentioned, gender, much like race, is also a construct. "'Body' is a key to an understanding of the roots of male dominance and the construction of female social identity," because, as Silvia Federici writes, "gender should not be considered a purely cultural reality, but should be treated as a specification of class relations."[6] In *Caliban and the Witch*, Federici's opus, she speaks to the subjugation of the biologically female body and how it's been exploited through the ages for the perpetuation and spoils of capitalism. "At the core of capitalism there is not only the symbiotic relation between waged contractual labor and enslavement but, together with it, the dialectics of accumulation and destruction of labor-power, for which women have paid the highest cost, with their bodies, their work, their lives."[7]

Capitalism is a cycle of continued fear, entrenched in a white supremacist standard where one should always want more than their neighbor and that they should compete, exploit, and steal if necessary to gain entrance into a mythical world, one where accruing money brings you true happiness. To want *more and more* is to be unsatiated, and thus to continue to live in fear. In *Braiding Sweetgrass*, Robin Wall Kimmerer speaks of the *windigo*, the legendary monster of the Anishinaabe people. "Windigos are not born, they are made. Its bite will transform victims into cannibals too."[8] Kimmerer explains that though this is a folkloric tale to scare children out of the perils of greed, there remains a profound parallel between mythology and reality—what is a *windigo* if not a modern capitalist? "The consumption driven mind-set masquerades as 'quality of life' but eats us from within. It is as if we've been invited to a feast, but the table is laid with food that nourishes only emptiness, the black hole of the stomach that never fills. We have unleashed a monster."[9]

Billionaires make no sense to me, but societally we allow it, we *accept* it as is. During the Covid-19 pandemic, we've had to digest the disgusting reality that rich people have gotten richer while millions around the world have died preventable deaths. Outside of the uprising for Black liberation, there was no real confrontation of capitalism. Is it too much to demand that a society that talks so freely about its sophistication and superiority might be the first to think about complete prevention by actually caring holistically (and equally) for its citizens? Let alone the world? As in wellness for all? If Jeff Bezos or Elon Musk or any and all billionaires decided to redistribute their wealth, we could unleash a true human revolution by redirecting wealth toward the gain of all. So, why must we encourage a society that enshrines greed and applauds it? What does it say about a society that is willing to live with this kind of discrepancy between classes?

With all this unearned capital, we could solve climate change. Why go to Mars when you have a chance of still healing this planet? A planet that has many of its citizens fighting to protect it. "When elites put themselves in the vanguard of social change it not only fails to make things better, but also serves to keep things as they are,"[10] Giridharadas shares. "Many of them believe that they are changing the world when they may instead—or also—be protecting a system that is at the root of the problems they wish to solve."[11]

◎ ◎ ◎

All my life I have been confused by the current state of the Muslim world. The Golden Age of Islam, which occurred alongside the Dark Ages in Europe (a term my father calls inherently racist, as it assumes because the West was in the dark, no other civilizations existed nor mattered), was a time of modern mathematics, philosophy, medicine, poetry, science, and enlightenment—but now, this same land is known for depletion, drones, and American-made or -backed wars, and thus chaos. Centuries-old empires, civilizations—suddenly vanished over mere decades, completely run into dust. As a child, I think this led to the dissolution of what history claims as truth. Under my father's accidental tutelage, I was taught about the Muslim world through VCR tapes he'd find at our local library or books from the university where he worked. This world, a time of Rumi and Shams, felt deeply aligned to the kinds of things that were beginning to attract me, deep histories of knowledge embedded in mythology and mystique.

For the longest time, if you were a marginalized person living in the West, you no doubt learned how to code switch. For me, this meant speaking in a different language (usually Bangla) to talk about other (namely white) people, especially when they were being rude or discriminatory, which growing up in Australia in the 1990s happened often. Because we weren't yet a society that allowed us to say "white people," this allowed for a silence to permeate about the obvious. White supremacy was rampant, and the devastation and ubiquity of white violence could be seen in every facet of life. Whether in the destruction and obliteration of my faith, or the formation of the country I was born in, or the nation-state I was raised in, or the land my parents are from but had to escape. Everything had been touched by white supremacy.

In March 2021, a white man shot down eight people in three spas in Atlanta, killing six Asian women. A law enforcement official said about the killer, "Yesterday was a really bad day for him, and this is what he did," as if being the gunman in a mass shooting is akin to running a red light. The deplorable nature of whiteness is that in order for it to be sustained, everyone has to go along with it. Too many of us have kept quiet about our living realities for so long because the fear was always . . . death.

In a piece about US military bases originally published in the *Nation*, Alice Slater reported, "In a series of panels over two days, conference speakers from every corner of the globe proceeded to describe the extraordinary cruelty and toxic lethality of U.S. foreign policy despite (Dr. Martin Luther) King's warning more than 50 years ago," at Riverside Church, where he declared the United States "the greatest purveyor of violence in the world today."[12] This, factually, is true. The United States has approximately 800 formal military bases in 80 countries, with an estimated 138,000 soldiers stationed around the globe. This is the same country that invaded Iraq under false presumption that Saddam Hussein, who was assassinated without trial by the American government, was carrying weapons of mass destruction. Who thinks about the wellness of Iraqis and Afghans after the American war machine? I can still hear Bush's colloquial drawl, sounding more like a Muppet than a president. Perhaps modern imperialism is unique to the United States, because its self-declared exceptionalism justifies imperial bullying, murder, and complete and utter destruction masked as solutions and diplomacy. The historian Roxanne Dunbar-Ortiz once calculated "the historical norm for the United States is to be at war."[13] Boy oh boy does the United States love to lecture, but it sure as hell doesn't care about the diligence it takes to actually be the society it declares itself to be. This sounds familiar, doesn't it?

"U.S. nationalism has a religious character. Its destructive mission is imagined as sacred," adds Slater.[14]

I'm grateful for the word *epistemicide*, because it's the only word I've ever seen that carries with it multitudes of deep forgotten history—the explanation of all that has been lost—and the subtleties of that. To live knowing your cultural knowledge has been stolen, exploited, robbed, or destroyed—while being told otherwise—to be taught of the West's majesty but not of its destruction is to be gaslit every day of your life. The colonized world, and thereby vast knowledge systems that were not Western, is the backbone of the modern world, including within the realm of wellness. The Muslim empire had sanitation and clean running water, and was also at the forefront of modern medicine, while Europe was in the Dark Ages. The vast Indigenous knowledge systems from Mesoamerica, Africa, and the rest of the colonized world were destroyed to ensure our health, our wellness, our livelihood, and our success could be hindered. "At the height of the Al-Andalus Empire in Europe, the city of Cordoba had a library of 500,000 books. This was at a time when other intellectual centers in Europe would have had libraries of between 5,000 and 10,000 books. The Spanish burned the library in Cordoba, and libraries elsewhere. They destroyed most of the codices in the Mayan, Inca and Aztec empires as well. Women's knowledge, which was largely oral, was simply silenced, as was the knowledge of Africa."[15] Federici adds to this by writing to the women who were killed in witch hunts across the globe: "[Witches are] the embodiment of a world of female subjects that capitalism had to destroy: the heretic, the healer, the disobedient wife, the woman who dared to live alone, the obeah woman who poisoned the master's food."[16]

◎ ◎ ◎

Everything is connected. Knowing and understanding that there is a very specific reason why the West is seen as superior comes down to the aggressive colonial PR machine that propagates the superiority of whiteness. Anyone who questions that is apprehended, quashed, and in many cases murdered by the powers that be. All we need to understand is that vast civilizations were toppled not only by being conquered, but also by being *silenced*. We were forced to forget so we wouldn't know who we are, what we come from, what our lineage, family, culture has taught us, history, language, stories, memories—stolen, robbed, put into museums—to make the West more money, as if we disappeared and went extinct. In *Unthinking Mastery*, Julietta Singh writes, "The colonized body embodies the histories of its oppression by recognizing in material ways that it is not free in relation to the world that surrounds it."[17]

But isn't our collective human reality the one we make, together? The more I think about the expansion of the body, especially within the context of evolution and how bodies have adapted, I see how much there is to uncover, and how many places we could go. Paulo Freire reminds us in *Pedagogy of the Oppressed* that the task of the oppressed is to liberate themselves because "the oppressors, who oppress, exploit, and rape by virtue of their power, cannot find in this power the strength to liberate either the oppressed or themselves."[18] A possible way of doing this, according to Saidiya Hartman, requires an excavation of the margins of history, meaning we need to collectively gather our histories that have been stolen, taken, in order to excavate the "ruins of the dismembered past" and revive the knowledge systems and practices that were deemed illegitimate, pagan, anti-Christian.[19] Reimagining history relies on objectivity within social relations, as well as an acknowledgment that all peoples have inherent value.

The optical illusion here is one of power, namely who has it and who doesn't, and this always comes back to the body. As mentioned through the annunciation of a technique like Taylorism, the way that bodies were used for capitalist gain is most prominently exemplified by the Atlantic slave trade. "The ruthless use of labor power and the extraction of profit are imagined as the consensual and rational exchange between owner and slave. This is accomplished by representing direct and primary forms of domination as coercive and consensual,"[20] Hartman writes in *Scenes of Subjection*, asking, "How is it possible to think 'agency' when the slave's very condition of being or social existence is defined as a state of determinate negation? In other words, what are the constituents of agency when one's social condition is defined by negation?"[21] How can we overlook such an imbalance of power? Facing our barbaric interconnected histories as a species means we must also acknowledge what Hartman calls the "brute force of the racial economic order,"[22] to fully understand how racial capitalism relies on anti-Blackness, and in general how dehumanization of a supposed willing subject—and, yet, paradoxically, a held captive—still coincides with the racial order of the world. We are still entrenched in these systems, and perhaps that's why we remain beholden to them. But the perversity of optimization, to assume that bodies need only be controlled to mimic machinery has meant again we accept things as status quo when throughout history we've seen that one absolute remains: systems and empires always fall. Part of that, I think, is because patriarchal rule is deficient. Men are inherently predictable, and absolute power corrupts absolutely.

The history of racial classification, or the formative racial optical illusion, starts with François Bernier, a French physician and traveler known as the first person to create a system of human categorization based on "race." In 1684, he anonymously wrote "Nouvelle division de la terre par les différentes espèces ou races qui l'habitent" (New Division of the Earth by the Different Species or Races of Man That Inhabit It), publishing it in *Journal des sçavans,* the earliest academic journal published in Europe, where he placed white Europeans as the norm from which other "races" deviated. Going so far as to distinguish four different races, where the Sami or the "Lappons" (Indigenous Finno-Ugric people inhabiting Sápmi, the northern parts of Norway, Sweden, Finland, and the Kola Peninsula) as the fourth and last race, describing them as "a small and short race with thick legs, wide shoulders, a short neck, and a face that I don't know how to describe, except that it's long, truly awful and seems reminiscent of a bear's face. I've only ever seen them twice in Danzig, but according to the portraits I've seen and from what I've heard from a number of people they're ugly animals."[23] According to Siep Stuurman, this "can be situated at the 'beginning' of the long and complex intellectual trajectory of modern racial thought." Stuurman also concludes, "The contours of the presumed races are in some cases ill-defined, the status of 'colour' is, to say the least, unclear, and Bernier's sketchy classificatory scheme is a far cry from the later, more theoretically-elaborated, racial typologies. What it has in common with later racial thought, however, is that it is exclusively based on physical criteria."[24] Stuurman adds in this chilling observation, "Taking into account that this was written shortly before the infamous *Code Noir,* regulating the Atlantic slave trade, was promulgated by Louis XIV, it is hard to avoid the conclusion that Black Africans are considered 'natural slaves' by Bernier. Elsewhere he speaks of certain cannibal tribes, like the Brazilians (the proverbial Tupinambá) and the Huron, in which nature has corrupted itself to such an extent that they 'retain less humanity.'" This reminds me of what Brazilian architect and urbanist Paulo Tavares writes in "The Political Nature of the Forest: A Botanic Archaeology of Genocide," in *The Word for World Is Still Forest*: "The fabrication of this epistemology was intimately connected to colonial imaginaries that functioned as one of the most powerful and enduring instruments in the historical extermination of Indigenous peoples."[25]

Bernier represents the violent colonial attitudes that determined much of the way colonizers related (or rather, didn't) to their subjects and, in particular, their bodies. Because he was the first to propose a system of racial classification that extended to all of humanity, his racial categories became the genesis of scientific racism.[26] Meaning it became the evidence on which modern science was then built, and this, of course, is deeply rooted in modern white supremacist rhetoric. In 1939, Heinrich Himmler, then the head of the Gestapo, praised two leading researchers at the Kaiser Wilhelm Institute for their work "that has contributed significantly to the scientific recognition of the racial parts of the national socialist view of the world." The regurgitation of racist rhetoric to murder and kill is not a modern invention either—the petri dish of genocide began with colonization, and we understand that no bodies are safe under this structure of violence that presupposes others' lives are more worthy *of life*.

In late 1658 onward until his execution, Bernier was the private physician for Prince Dara Shikoh, the great Mughal emperor Shah Jahan's son, and then for Dara's brother and rival, Aurangzeb. In India he was able to confirm many of his findings because "he chose not just to describe men and women from various locales in India but to sort them based on their skin color," Strings writes in *Fearing the Black Body*. "In fact, he imagines himself an astute interpreter of existing social categories in India, in which he sees skin color."[27] As if seeing color, and the inherent supposed superiority or inferiority in that necessitates an abuse of power—as if claiming domination was all it took to be so.

In an interview with Vinson Cunningham for the *LA Times,* scholar Robin D. G. Kelley speaks to racial capitalism in particular, and why difference encourages exploitation: "So, for example, property may be capital, in the Marxist sense, but property *values* are dependent on things that are nonmaterial—that are ideological, or super structural—like race. Capitalism is rooted in a civilization that is based on difference. This doesn't at all mean that white people are the enemy, or that Black people are all victims, which I totally reject. It doesn't mean that all white people benefit. It just simply means that capitalism is structured through difference."[28] That's one point I keep thinking back to with this research. So much about understanding the body, our collective bodies, but also our individual ones, is about understanding how society envisages us. Because all of this information is stored in our collective experience, our histories that the entire working of the world is determined upon. A refusal to see how this is not only all connected, but how this influences our future, is something that we have to recognize. There is no future that isn't tied to reckoning with our past, and we must challenge those who choose to deny their responsibility.

◎ ◎ ◎

We wonder how physical disease plots the body—where do these things derive from? Why don't we understand that the mind is not only connected to the body, but it's linked—they are two different sides of the same coin. Thoth, the Egyptian god symbolized by an ibis, later returned as Hermes in Greek mythology, seen as the father of alchemy, and was often represented by two snakes intertwining to represent healing.* Ancient symbolism, with the backing of ancient cosmologies understood to heal, meant facing two fronts. Trauma isn't always the stored knowledge of what has been done to you; sometimes trauma is about what you've done, the mistakes you or your ancestors have made, the grave things they were incapable of metabolizing that now sits in your body, like a deluge, bubbling beneath the surface. We have to begin to understand the spectrums of chronic pain, fibromyalgia, arthritis—and how they are tied to trauma. As ancient civilizations have stated, there is spiritual law, and it's not as fickle as human law. Disease lives and is carried through families, trauma exists in ways we can't determine, like hungry ghosts planted in the body, and one day they have to come out. Sooner or later, they need to be exorcized. Isn't it interesting that certain familial genetics hold cancer, while others don't? What I'm getting at is that our bodies are flashlights toward our ancestors; to better know them, we can know ourselves. We can trace what they are speaking through our bodies, but we must have the patience to listen. I believe if this work does not happen, it will continue to unfurl and cause pain. What I'm trying to say is that time is running out.

I would argue that science has also shielded and provided a

* "At a cosmic level, the Egyptian God Thoth (Tehuiti) symbolized the mind and memory of the Demiurge (Creator) and at the human level he was connected with the various faculties of the mind, its discerning as well as its analytical qualities. Thoth was the patron of magic because he was the embodiment of intelligence (human and divine), while his Greek counterpart, Hermes, the messenger of the Gods, symbolized the 'dimension' of the mind as the intermediate level between the human and the Divine. Hermes also came to represent those intellectual and eclectic qualities (in the human being) that were characteristic of the Hellenic spirit and that allowed the birth of the Hermetic tradition." Source: https://library.acropolis.org/the-magic-of-thoth-hermes/.

gateway for more violence to occur through the lack of objectivity, through an obsession toward control. We've lost sight of ourselves and thus each other. But it's important to see how bodies are political. Our only path is toward an authenticity that holds all truths. Yours, mine, and everybody else's on this great and vast Earth.

CHAPTER 6

ON BODY DYSMORPHIA

It's taken me a long time to articulate that I'm an incest survivor. When you face the kind of trauma I have in the just-over-thirty years of life, some things begin to blur and other things are accepted as just the way things are. I cannot exaggerate how long I've felt as if, though terrible, my life had *happened*, the past had passed, and I was desperately trying to look forward, to forge a new life. Maybe the reason I'm so obsessed with looking at everything is because I spent so much time in the shadows of my life, in the back seat, believing I was not in control of what happened to me.

It's why I moved to New York, leaving my life in Australia far behind me. When people would admire my bravery, I brushed over the details of my past, believing that the mirage I was trying to build would be impenetrable, and that the life I was creating would stand in for my former life. Sooner or later, I'd be so far from it that it would never be able to catch up to me. I was running away from home, but quietly. I didn't want to alarm my mother but wanted to go far enough so she couldn't bring me back. I felt as if I was abandoning my family, but I hid from those ugly feelings, and my rhetoric of failure festered into guilt.

I guess I didn't know what was happening to me as a child because there was never a reprieve from it. Now, after a few years of therapy, I can trace and see how on top of all of this, the impacts of her physical abuse and the sado-sexual parts of the abuse were never something I interrogated. Everyone is groomed by their aggressor to rationalize the situation. So, as a child you learn to fight against yourself. I was taught a narrative that vulnerability is a weakness, that acknowledging trauma is unconscionable. You live with these things and the cycle continues. But that wasn't how I wanted my life to turn out, so when I finally got far away from the miasma of my past, I started at the very beginning by asking questions like *Why? Why did I have this fucked-up life?*

My body was never mine to begin with. When I first started to question what Maté describes as "civil war inside the body," the memories I had hidden, the dirty secrets stuffed inside the folds of my skin, began to appear like apparitions. I believe they came as I was ready. I had just ended the most long-term and significant relationship of my life, and after what I had deemed a "failure" or "failing," I think psychologically a part of me broke. This conception of myself that I'd been holding on to, this person who was "together" or "had answers" no longer seemed true. So much of who I am was shrouded in fear because I thought the real me was too much, too loud, too everything. To step into your power when you've never been allowed to is an exhausting struggle, because the rewiring is required at such a formative, and sometimes inaccessible, part of yourself. I've had to connect to a lost self and send her a signal after keeping her trapped in a hole underground for so many years, denying her the right to even speak, to even voice her perspective to me. Whenever she would get close, I would scream like I was being sucked up by a Dementor. To revisit painful memories made me extremely scared.

I've been dissociating for longer than I can remember, and for most of my life I actually thought disassociation was normal. I assumed I must not know how to be a person, and because nobody was teaching me or taking me into account, that's when I began to dream. In that world, I just was. Accepted. Not a body, or a person, or a kid. Confused by my function as a person and already feeling deep existential dread by the age of ten, I decided that I would learn how to serve others, while fantasizing about killing myself. I thought to martyr myself would give me the love I deserved, but also the recognition of keeping other people's secrets in my body, the way I kept quiet, how my body took punishment. There was a part of me that wanted both, complete destruction but also love, a partner, a family. It's actually what I have wanted the most—people to love me, all of me, wholly.

For the majority of my life, I was considerate, I was kind, I was organizing at school and excelling at leadership. At home, I was desperately afraid of punishment, and got enough of it already, so I tried very hard to be whatever anyone needed of me. As a lot of my mother's abuse over me was possession—possessiveness over my body and personhood—I wasn't allowed out of the house; in fact, nobody was. So, we all stayed in the perimeters, always within my mother's reach, always there to satiate some untenable want. I felt watched when she was around and surveilled at a distance when she wasn't near. Her presence was a void and felt vampiric but overly critical in nature, so I was always expecting judgment. All of this made me extremely, exhaustingly hypervigilant.

As I got older, I began to seek out people who reminded me of my mother, finding comfort in the familiarity. Until only very recently I thought of it as a result of an astrological placement, as I was ruled by Saturn, which I knew meant I liked challenge, hard work, and difficulty as a means for evolution. It took me a while to understand it didn't have to be this way, that I had agency to restructure a new dynamic with people who I presumed loved me. Even still, because I was conditioned to live in fear at all times, my imagination didn't expand to the possibility that I could have a peaceful life. One way or another, we perpetuate trauma, many of us even pursue it, because we fall into its cyclical nature. No surprises, but duh, *conditioning conditions you.* This also comes back to limiting beliefs, as well as policing the imagination. Trauma is the link. If you cannot fathom being happy, do you think you're ever going to *really* get it, *be it?* The more you deny yourself joy, by not even believing in its possibility, the more you encourage the *lack of it.* This might seem to those in the rational camps as a form of toxic positivity, but I don't know what to tell you—the moment I stopped victimizing myself, something shifted.

@@@

I distinctly remember when my gender dysphoria began. Around ten, after my father's first trip to Malaysia, I received Adidas tracksuit knockoffs, and like a Tenenbaum, I wore them everywhere. It was the first item of clothing that I felt was made for my body. Nothing was too tight over my suddenly voluptuous physique. I hated my feminine aspects, always have. At the time I was playing a lot of sports as a way to offset my ugliness, and, I assume, to give myself and my body some use.

We were living in ultra-white 1990s suburbia Brisbane. I constantly felt like the odd one out, so I would lie about my life to fit in. I told everyone that my parents made $500,000 a year and drove BMWs. It was a kind of wealth I knew bought respect, and I wanted to have a reason to be respected. If it wasn't for my presence or personality, it would be for the lies I told. I was always performing, always a little class clown trying to make everyone like me. Then, one day, this was all exacerbated. On the soccer field, just before a game, I was sexually assaulted by another girl my age. She touched me in broad daylight, in front of everyone, but nobody said anything. No one told her to stop, no one told her it wasn't her body. Nobody, again, protected me.

I was used to this, of course, but I was haunted by her actions. By the casualness with which she touched my breasts, my un-bra-ed ten-year-old breasts. There was also another layer of the violence—the girl was white. Not only did I understand that my voice next to hers would be negligible, I also understood, on some level, *her* whiteness would win. This is why racial imagination is so devastating, it becomes ubiquitous. Someone like me, with the life that had made me believe in my unworthiness, was doubly broken, because, on top of that, I had to unlearn whiteness and my position as a nonconforming Muslim Bangladeshi *person*. Even at a young age, I never bought into my body as a "female" body. To me, it was such a disgusting entity, one that quaked with desire and impulse and hunger. But that part of me, the oversexualized part, made me feel sinful and remorseful and confused by my own proclivities. I didn't understand why Allah had made me this way—why was I made into such a wretched and horny beast? Back then, though my body's template was so hard to understand, I knew I must be demonic. My mother always told me I was, and I had a body that continued to disappoint me.

Around seven to ten I was gaining weight very fast, eating any and all the junk food that I could get my hands on. Food became the comfort I didn't have in my actual life, but the chubbier I got, the more my disdain for my body grew. I was terrified of being looked at but depressed when I wasn't. I didn't understand my sexuality either—I was crazy about boys but also fantasized about girls. The more I dwelled in this confusion, the harder it got for me to understand what was *mine*. The more I let others determine who I was for me, the harder it felt to ground myself in who I knew myself to be. I was also fighting my mother's voice in my head, the one that would call me names and remark at my stupidity, insolence, and demonic qualities. I never understood what I was doing wrong, so I would go into the cycle of showing as little emotion as I could in front of her, especially when she was in a punishing mood, and I would just shut myself off, disassociate, and go into space, where I felt most safe. Where I could handle anything she threw at me with a hidden disdain.

It was around this time that I began to realize that my own masculinity might protect me, but this was also aided in a confusing distillation of how my body was read in the context of whiteness, a prop—yes, something to touch at will—but also an *other*, an unknown entity. "In my teens, I thought myself to be literally less perceptible than others," Julietta Singh writes in *No Archive Will Restore You*. "I believed that my body was less detectable than those white figures everywhere around me."[1] It doesn't always come down to some of us "wanting to be white," it's not as simple as that, because in that statement the context of white supremacy is evaded. For many of us, especially from South Asia, where colonization was executed differently and the impacts of partition have truly *never* been faced, a lot of us in the diaspora have been misinformed about who we are. I don't know why culturally it's more honorable to *not* speak of what ails you, but the more you do that generationally, the more disconnected you become from the whole truth, from the pain at the root, and thus the more unconscious you become. When you don't know who you are as a person, you seek direction, emulation. Whiteness is a catch-all because it convinces you that if you participate in it then you, too, can have a good life, aka a white one.

It's taken me a long time to understand the totality of the violence whiteness wreaks. White supremacy is a complicated beast with many arms, and that, in essence, is an important facet of why and how it's propagated. The roots are deeply embedded in modern society. In my twenties, looking back at my past, I was judgmental of my choices as a child. As I'm prone to think I am always at fault, I assumed that I just wanted to be white, but really what I wanted—I now understand—was to be accepted, to be "normal" as it was sold to me. "Often, we don't appreciate our bodies because we don't feel them," Selassie writes in *You Belong*. "We have ideas about the body, but we don't feel them. We have ideas about the body, but we don't actually sense our bodies. If we can't sense the body, we can't belong to it."[2] If you aren't adjacent to what is read in the dominant society as power, then naturally you are isolated. This happens in so many ways—if you are disabled, if you are working class, if you were sexually abused, if you have a mental illness, and, of course, if you are not white. This festers a unique sense of isolation that Selassie articulates as a lack of belonging. Many of us, I imagine, feel as if we don't belong in our own bodies, and thus the world at large. This formation is perhaps something we gathered from our childhood, and it might not even be something that's ever articulated in your mind, and yet it's an unsettling feeling of knowing that you are not accepted, and may never be accepted in your totality. This is what brews unworthiness.

"It is intuitively easy to understand why abuse, trauma, or extreme neglect in childhood would have negative consequences," Maté writes. "But why do many people develop stress-related illness without having been abused or traumatized? These personas suffer not because something negative was inflicted on them but because something positive was withheld."[3] When you forget, your mind might turn one part of your brain off, but it never really forgets, just how your body never does, either. Some of my literary idols spoke to the lovelessness or hardship of their childhoods—Susan Sontag, June Jordan, Audre Lorde. All of them died of cancer, all of them wrote extensively about grief.

The wildest thing I've learned about abuse is that it's like an energy, a current that runs through you that you can pass down from generation to generation. It's known as a cycle of abuse because it's an energy that is recycled. The more I untether from my childhood, the more I can make sense of it, but it requires my full attention, actually looking at what's there, and no longer denying what my conscious mind and my body know to be true. It takes true acknowledgment to heal. And then it takes action. Those two things, like Hermes wielding alchemy, are healing.

I had caught wind of my mother's life from a young age, her life before us. My sister would tell me in whispers about the locked attic my mother was forced to live in for four years. That's when I began to realize that there was something deeper there. From a young age, as her confidante and her keeper of secrets, I'd spend many nights lying by her side and massaging her body. It was generally the only touch she allowed from me—though my mother was a bully and believed she could touch me wherever she wanted, I was not allowed the same access. Nothing about that seemed odd to me, but the quick way she would react even to a hug, or affection, from me or my sister made me realize that there were many loose threads waiting to unravel, to be looked at. Because the power she exerted over us was through our silencing, we weren't allowed to state the obvious or even ask questions. It's as if the older we got, the more domineering she became. She could shut us down with episodic anger that was rapturous, but in smaller ways, too—by ransacking our rooms, listening in on every phone call, watching us like a hawk, beating us if we were home later than our expected arrival time. The more she revolted, the more I divorced from reality. In order to accept my life, we all had to keep the charade churning. We all took our parts in the play of my mother's drama.

It took me years to understand that she and I did not have a normal relationship. Outside of the abuse she wielded on the three of us quite equally, there was something sinister in the way I was beckoned by her. I don't think she was purposefully exploitative, I think we were both longing for some tenderness, and for many years I blamed myself for that, for wanting care, for participating in my abuse, for never saying no. For allowing myself to become hers, physically. To be owned by her. Possession is sexual and exploitative because it always requires someone's submission. When I think back to a younger version of myself now, trying to understand how I just handed myself over, I am speechless. Yet this is also who I've always been, who I've always been forced to become: a girl at the behest of her mother, doing whatever I could for validation and concern. The saddest part is how this state of relationship—me as the forever captive forced to serve her captor—has bled into most of the intimate interpersonal relationships I've had.

The lines between responsibility and codependency often become blurred in abusive dynamics, as do the lines of balance that allow for both parties to have a say in any relationship. I seem to attract relationships where I serve people, endlessly. It's what I think will bring me love. But within these dynamics the relationship always implodes the moment I stand up for myself after complying for too long. Because of my childhood, not only did I not know how to stand up for myself or understand it's important to have your own boundaries, your own standards—I also didn't know how to show anger. "The person who does not feel or express 'negative' emotion will be isolated even if surrounded by friends, because his real self is not seen. The sense of hopelessness follows from the chronic inability to be true to oneself on the deepest level. And hopelessness leads to helplessness, since nothing one can do is perceived as making any difference."[4] When you learn to accept unlovability, you also have to accept everything that comes with it.

However, almost like a blessing in disguise, the brutality of my life has been most revealed to me in the breakdown of some of my closest interpersonal relationships. I've started to accept that most people, even if they say they love you, will take advantage of you if you let them. I've seen people I thought loved me become entitled to me—especially in moments where I've *finally* enforced boundaries. It's when I become the bad guy, the narcissist, the one who's in the wrong. As Maté writes, "A pathologically enmeshed family system is characterized by a high degree of responsiveness and involvement. This can be seen in the interdependence of relationships, intrusions on personal boundaries, poorly differentiated perception of self and of other family members, and weak . . . boundaries."[5] It's alarming how many of us perpetuate the family dynamics we learned as children, how many of us choose enmeshment as a proof of intimacy.

After three decades of dealing with my mother's drama, I finally developed the nerve endings to ensure my safety when I see that energy coming my way now. These days, I want ease, I want softness, gentleness, care, and compassion. At the same time, the older I get, the more I know it's necessary for me to be surrounded by people who get the full scope of what I'm working with, that they understand my past in order to get me, and hold me, at both my present and future.

◎◎◎

We have to learn to have a better understanding and discourse when it comes to bodies, and understand how things impact us as well as what it means to understand what your body is telling you. "This may seem odd," wrote Hans Selye, "you may feel that there is no conceivable relationship between the behavior of our cells, for instance in inflammation, and our conduct in everyday life. I do not agree."[6]

As a child, perhaps through signals from my mother, I began to understand that I couldn't be my full self, so I had to hide and dim parts of who I was in order to be acceptable, and therefore safe, at home. Otherwise she'd hurt me, or berate me, and to save us both some time, I began to shut down parts of myself, only showing the side that was happy, neutral. As my body began to blossom, I was monitored even more closely. Any sense of sensuality or even sexuality was quashed because subliminally, my mother suggested everyone was out to rape me, including my father.

I existed in that fear for most of my life, a fear that my beauty would be used against me, so hiding was the best choice. But the irony is that Baby Fa—the small, child part of me who doesn't care what her body looks like—she wants to be witnessed, she wants to be seen, she wants to come alive and play and be innocent and do all the things that she was never allowed to do and have a childhood. I have a lot of grief over being robbed of my early life, and so I have a lot of anger toward people who have had good childhoods. It's a privilege, I'm not sure why we don't understand this, especially when the difference between having a good childhood and not has dangerous long-term side effects. Maté articulates it perfectly: "Infants whose caregivers were too stressed, for whatever reason, to give them the necessary attunement contact will grow up with the chronic tendency to feel alone with their emotions, to have a sense—rightly or wrongly—that no one can share how they feel, that no one can 'understand.'"[7] The way we downplay this reality results in a collective gaslighting. Despite the odds against us, children of neglect are forced to live normal lives. We are expected to get on with it. If we speak about our pain, we become burdens, or worse, victims who won't shut up. But I'm speaking up for my kin, for the children like me. I'm making the connections and the conclusions because it's time we face these things, too.

My body feels like it's been thrown around for generations, though now it's finally in my grasp. It's important for me to express how I feel after being silent (and silenced) for so long. To be exotified, fetishized, since childhood is to lose yourself. By six, my mother would tell me to close my legs. "Who are you trying to impress?" she'd ask me. The disgust that I would feel for my tiny body was unimaginable, like she was pouring liquid poison down my spine. This was her favored choice of harassment, making lewd sexual jokes. Talking about our bodies like we had adult agency, forever blurring the lines between mother and captor. If I reacted poorly, she would accuse me of being dramatic, an evil child who was too sensitive. In fact, that was a regular attack on me, that I was too sensitive and emotional. Sometimes this would turn into a bad mood, which would turn into a days-long silencing battle, or of banging pots and pans, walking around the house enraged. Needless to say, the best way to move through her uncomfortable groping, touching, poking, prodding, or pinching was to go along with it. To nod your head and appease. But the more you did that, the more you handed yourself over to her.

There are so many layers I'm still uncovering. There's so much that I barely understand about the things that have happened to me, but I do understand now that my questions about my gender were rooted in my fear that if I presented as being more femme that I would be faced with danger. On top of that, to neutralize my body in my father's presence, so not to be raped, I became his son. I became a dutiful, intellectual son who spoke to him about politics and the perils of capitalism, and I, again, hid the other parts of myself. The sexual parts, the feminine parts. I always felt like a stranger inside of myself, and I grieve never being given the chance to be truly comfortable in my own skin, the chance to figure out my nonconformity on my own. I grieve having made choices out of fear.

In his fascination with classifying race, Bernier had a more sinister motivation and fascination, Stuurman writes: "By focusing on the aesthetics of the female body to the exclusion of all other criteria, the masculine, sexualized gaze naturally fits in with a discourse on race which posits physical, biologically determined differences in looks, colour and bodily shape as the ultimate foundation of a 'new division of the earth.'"[8] Stuurman goes on to explain that the fetishistic and upsetting ways in which Bernier speaks about women from Persia, Kashmir, and Circassia was claimed as objectivity, because men like Bernier, simply because of their whiteness, were seen as having verifiable opinions, merely detailing what they see, when what they could only ever see was the absence of what they were, they being the default and the litmus against everyone else.

The ways in which women's bodies have been fetishized and weaponized is an endless journey of peril. We live in a fatphobic and ableist world that assesses beauty in a completely unrealistic sense, and yet it determines people's livelihoods. I've spent too much of my life caring what people think about how I look and not enough time genuinely taking myself in. Because I've spent the majority of my adolescence to adulthood living with body dysmorphia and dysphoria, I spent so much time hating myself for that confusion, for not having any answers for why I felt the way I did, and for not feeling normal because of it. Up until my mid-twenties, I was afraid I was on the verge of losing my sanity, like my mother. Afraid that I'd wake up one morning having completely lost myself through the night, forgetting who I was entirely. Nothing was in my control or in my domain, and yet I survived. Humans are strange like that.

I was not raised to think my body was mine, but my mother's first, my family's second, my culture's third, and my faith's fourth; never mine. Sexual abuse at the hands of the same parent that surveils you creates a sense of forever fear; nowhere is safe. I had to travel thousands of miles from the site of the violence to experience some form of safety. I always felt hunted by her, possessed by her. This psychosexual explanation for the blurring of parent and child can be explained through the theory of enmeshment,* a concept in psychology introduced by Salvador Minuchin to "describe families where personal boundaries are diffused, sub-systems undifferentiated, and over-concern for others leads to a loss of autonomous development,"[9] which sometimes mimics a lost or absent spouse for the parent. This requires a certain codependency to continue, but these

* Enmeshment is a theory within psychology whereby indistinct boundaries in familial relationships lead to the lack of independence and autonomy within a member's development, typically a child.

effects are devastating to the child. In her essay "Situated Knowledges," Donna Haraway explains further, "We are not immediately present to ourselves," a lasting legacy "especially true for survivors of trauma and for people who have generations of trauma history, such as the traumas of alcoholism, abuse, war, and colonization."[10] I used to think that trauma and abuse were all relative, but the older I've gotten, and the more I've begun to understand the real tenor of the violence of experience is exactly why I have felt so isolated in this world. Not everyone can meet me here, and not everyone wants to. I've gained language to understand the intricacies. When I talk about my life, I can instantly see who understands this, and there are very few who *really* get it. Those who do usually, also, have been betrayed by a parent. Generally, if you had a healthy relationship with your family, the different paradigms results in a lack of computing. It takes a similar action as a white person trying to understand race—it requires humility and an understanding that you might not get it, but that it's not yours *to get*. When it comes to abuse, everyone is suddenly the judge, jury, and executioner. But the clearer I get about my own, the easier it gets to end relationships where my full dimensions are not considered, especially with regard to everything that I am. This means prioritizing the realities of what happened to me, even as I am trying to heal from them. I'm speaking to an uncomfortable truth, and I no longer wish to demean myself by lessening the pain for others and their comfort. I understand that my own testimony is what is most important, for *myself*.

A few days after the shooting at the Gold Spa in Atlanta, I went for a massage with my regular therapist. The night before I had finished watching *Allen v. Farrow*, the HBO series about Dylan Farrow's abuse at the hands of Woody Allen. Throughout the four-part documentary, I was able to understand Dylan clearly because I was actively confronting similar realities in my own life and had made similar connections. The footage of Dylan speaking about breaking up with her boyfriend as a teenager, the way she held herself in her body, her trajectory from this open, bright child into an uncomfortable sad girl—and even more isolated adult—felt like my own. Much like witnessing James Safechuck *be* (not even necessarily speak, just watching his body language) reminded me so much of the hesitation that exists inside of me.

Feeling vulnerable, from life and the recent shootings, I was determined to get some relief from my regular massage therapist, whom I had been working with for years, and who was aware that my childhood sexual trauma was the root of so much of my chronic body pain. That day, halfway through my massage appointment, they started talking about the Farrow documentary that they had also been watching. Even though I felt an initial discomfort at the choice of the topic, I trusted this massage therapist and expected them to know my boundaries. That maybe ass naked on a massage table was not the place to hear about child sexual abuse. And yet, inevitably, despite my hope, the conversation moved in a direction I had not been prepared for, one where my massage therapist claimed that Dylan was lying. That, apparently, according to them, the whole thing was staged as a way for Mia Farrow to get back at Allen. Lying on my stomach, completely naked, I became enraged. I came to rest, to find safety. If not here, somewhere I pay to feel rested, where were survivors' bodies actually safe?

I don't think people who aren't confronting their traumas fully understand how difficult the work is mentally *as well as physically*.

The more you're able to face it, the harder it becomes to ignore *anything* as your sensory capacity becomes more and more precise. As I lay there on the massage table, a few things ran through my mind: I was on my stomach, much like how Dylan was on her stomach when Allen molested her. How I was so many times proverbially and otherwise with my mother. Playing dead, playing dumb, thinking of the sound outside. Then, I thought of the women slaughtered by a white man with a "sex addiction," and how Asian women's bodies have been sexualized, fetishized, and exoticized throughout time. I thought about how many times my body has been compromised and how that was again happening, but in a different way. These days, the more I'm in my body, the more I react *with* it. So, as my massage therapist continued talking, I was able to locate the anger and express it. The session came to an end a little while later, but the damage had been done.

I cried all the way home on the train, grateful that I was wearing a face mask that helped hide my uncontrollable weeping. I wanted to scream. It felt like something had crawled up from underneath me and lit me on fire from the inside. When you have been abused and you know how to feel into it, this screaming void, you can feel all the souls with you that carry this pain, and the totality of it is almost too much to bear. As soon as I got home, I called my friend, also an incest survivor. She understood, she listened.

Women and femme bodies are extremely vulnerable in this world. But it's important to note that many survivors, especially children, are also cis male and masc identified folks. There is no binary between who is abused, and by whom (the abusers are, in most cases, cis men, but, like in my own life, women and non-binary folks are also perpetrators of sexual violence and rape), so, these days, I'm unconcerned for those who cannot hold the multiplicity of experience within the realm of sexual abuse. There is no universal reaction by a survivor, and survivorship doesn't look a specific way. We give such little space for the perspective of someone who has been apprehended at this level and allow others so much space to "debate" these intricacies on our behalf. But if we want to stop this kind of abuse from happening, a mass societal shift is needed. This is not within the spaces of science or law—this is an issue of power.

In her work, Megan Boler argues that a "pedagogy of discomfort" might be ethically imperative for us as a species. "Practicing and teaching our discomforts can become acts of learning to live with the ambiguities and uncertainties of our complex ethical entanglements," writes Julietta Singh.[11] Allowing conversations about discomfort helps us shape actual places of safety—by recognizing our physical and mental realities. Why are we so unwilling to hear out those who are speaking their truth? We silence the stories that threaten sexual hierarchies, as if women are only out to get men, when in fact society has been created to benefit men, and all the evidence is there, blasting like a song that won't shut off. This exists even in what is seen as a privileged or prioritized body. In *Fearing the Black Body,* Sabrina Strings writes, "The phobia about fatness and the preference for thinness have not, principally or historically, been about health. Instead, they have been one way the body has been used to craft and legitimate race, sex, and class hierarchies."[12] Everything is also about power.

As we begin to crumble our psyche and the specific ways in which we regard our sensibilities, we must also look at what has shaped our view of our own bodies and others. Reflecting on the higher rates of hypertension in Black communities in the United States and autoimmune diseases and rheumatoid arthritis in South Africa, Maté writes, "the major factor would seem to be the psychological pressures of living in an environment where official racism directly and overtly deprived people of autonomy and dignity, while it uprooted people from their traditional family and social supports."[13] In a story about the Bedouins, Maté shares, "ulcerative colitis has been noted after settlement in Kuwait City, presumably as a consequence of urbanization . . . Explaining that the drive for 'globalization' has undermined the family structure 'to tear asunder the connections that used to provide human beings with a sense of meaning and belonging.'"[14]

Embodiment is political, especially when many of us have been systematically disembodied on multiple fronts. To find peace within chaos is an act of resistance, to accept the chaos, to move with it, to return home to a body that has never felt like yours requires so much patience that at times it almost feels easier to give up. So, you fall into the depths of that narrative again, that nobody understands or believes you, and that you are alone. Then, one day, you remember and are able to sit in stillness with your body, you arrive with a sigh of relief and sadness at how easy it is to forget yourself. Here's a reminder that it's an off and on journey. It's not linear, but each day can be an act of becoming.

In Adrienne Maree Brown's *Pleasure Activism,* my friend Amita Swadhin writes: "It's hard to talk about pleasure when most of your life before leaving home at seventeen is a careful balancing act: hide the trauma, hide the truth, learn how to pass in the world as a 'normal' kid, a kid who isn't being tortured at home." It takes a lot "to learn there is a difference between hedonism that enables dissociation and disconnection versus joy and pleasure that enable presence and intimacy."[15] Sometimes one's human journey feels and looks crunchy, sometimes it takes a lot of time to be "human," but these days it becomes clearer to me that we just need to redefine what that means for us. And that we can do that societally, too. Clearly these labels and archetypes aren't working. Inevitably, so much gets lost when we are determined by measures that weren't written to house anyone's differences.

It's time for reinvention, it's time for dynamic change. If the way humans engage with each other—and therefore our external bodies—is inherently prescribed by colonial, patriarchal, and white supremacist divide (on top of the relations between sexuality, gender, and class), then this entire infrastructure needs a rebuild. Audre Lorde reminds us: "Tomorrow belongs to those of us who conceive of it as belonging to everyone, who lend the best of ourselves to it, and with joy."[16]

◎ ◎ ◎

Every day I'm starting again. That feels the most true for me these days, acknowledging that. Body dysmorphia is a complicated assignment. It shifts, like we shift, with the moon and the tides. Some days I'm happy, some days I can look at myself and see beauty. Many years ago, my trauma therapist asked me what I liked about my body. I stumbled on the question at the time, so she gave me an exercise to observe and be with my body. Many days, this work feels like engineering a connection where there was formally none, which is why I spoke to the mind first. Healing the mind is a very important component to healing the body, because in order to actually have lasting results, you have to work in tandem with both parts. There's a larger metaphor here.

Today I take solace in the words of Adrienne Maree Brown: "I touch my own skin, and it tells me that before there was any harm, there was a miracle. I confess daily, 'Here is where it hurts.' I let the healing come through connection. When I feel like a failure, I look at my plants, at how they wilt and seem to be dying, and then water and sun and my loving words bring them back to vibrancy. I let water move over me, the sun change me, love reach me. I root down into the soil and back into my lineage, which reminds me that everything is temporary, but nothing disappears, this is how life is."[17] Surviving is a maelstrom, but there's nothing else I'd rather be doing. That's gotta count for something, right? If coming back to myself, and finding love here, isn't a miracle, then what is?

CHAPTER 7

ON WHITE PEOPLE CO-OPTING YOGA

I have been practicing yoga on and off since the age of thirteen, more than half my life. It was my entry point into the wellness world, outside of my sister. Like a calling—maybe an ancestral longing that I've had my entire life—as a child capsized in diasporic ambivalence, yoga and meditation felt like portals into a lost, ancient self.

Perhaps it was also the closest tangible understanding that I had to being South Asian. The impact of partition, colonization, and the liberation war of 1971 had created a deep void of understanding in my parents. I've felt their longing to be known, and their longing for a homeland, or self, or family that never existed in the first place. Dislocated from a sense of who they are, my parents' unknowing of themselves was passed down to me, fed into an already unstable sense of self I had of being away from my homeland, in spaces that couldn't comprehend what it meant to be Bangladeshi or Muslim. I carried my shame and my parents' shame. I carried my mother's shame for the violence she had experienced in her body, that she took out on my body, and all the unprocessed feelings she had of herself that were funneled into me. I carry the war my parents survived and the violence they and my entire family had to survive.

These things don't disappear within a family—they linger and then they can very easily turn into disease.

My mother was a terrible parent, though I don't know how much of it is truly her fault. When you don't have the resources to heal, how do you? In *My Grandmother's Hands,* Resmaa Menakem speaks to this through the concept of reenactment: "This may seem crazy or neurotic to the cognitive mind, but there is bodily wisdom behind it. By reenacting such a situation, the person also creates an opportunity to complete whatever action got thwarted or overridden."[1] I don't know if my mother was ever that hyperaware—her moods were dicey and her cuts were like the hooks of a machete. Sometimes it felt like she was hacking my heart, wanting a response. I would never succumb, I would take whatever physical violence I needed to take to walk another day toward my freedom, one I was plotting. But my mother always felt like she was fighting with her internal ghosts—the *jinn,* she fears, that haunt her.

I was born into a generation where marriage wasn't forced on me in the way it has been for so many women before me, especially those of my cultural and religious background. I think this tendency to force women into marriage is a patriarchal flaw found across the world, and it's a way of controlling women, by setting their value based on male perception of their worthiness. My mother was never given an understanding of her own worth, and I have always felt that I was feeding her most of my life force, my energy, my vitality. I don't know much about my maternal grandmother, but I do know she was an orphan. When my mother would describe her, she would explain that she was someone who wasn't emotional. I can see how that lack of felt love transmuted into my mother's feelings of unworthiness. These patterns always find a way in the family like an untenable weed with roots that are obscured by the depth of the

Earth. There are so many demons in my mother's psyche that I don't understand, but as I try to uncover her, to see her clearly, I discover how much has been shrouded either in the history of my family or by the passing of time.

Out of my whole life, I've spent about two years in my parents' homeland, Bangladesh. I can speak the language, read and write it, and yet when my sister and I have asked for more, of understanding the lineage or culture, nothing was ever known or given. I've asked my father to archive his memories, and he does, poetic in his own way (he spent the entire time of the pandemic's first year by translating two hundred never-translated-before Rabindranath Tagore poems), and he writes about his memories, slowly, but I can see how they're like pushing blood out of a stone. Recently he told me his family moved to Bangladesh by accident, and that one entire part of his father's family lives in India. Partition ruled the lines in the sand, and in 1947, my paternal grandfather was working for the postal service in what was to become Bangladesh. He decided to stay in West Bengal and not return to India, to his family, out of obligation to his work. I am one of the many children of the diaspora who don't have any answers to the many lines that connect and divide their familial past, and my life's work is decoding all of it.

This comes back to yoga only because there's a responsibility (in literally the act and understanding of yoga) to hold multiple truths. "No man is to be judged by the mere nature of his duties, but all should be judged by the manner and the spirit in which they perform them," writes Swami Vivekananda in *Karma Yoga*. "We shall find that the goal of duty, either from the standpoint of ethics or of love, is the same as in all the other yogas, namely, to attenuate the lower self so that the Higher Self may shine forth, and to lessen the frittering away of energies on the lower plane of existence so that the soul may manifest them on the higher planes."[2] To teach yoga in the West, especially if you are not of South Asian heritage, by definition means you are appropriating it. However, the conversation of yoga rarely furthers the dialogue of how do we deal with the mass co-opting of Indian culture? Especially when yoga itself is unknown to so many of the people it belongs to.

The continuation of teaching a sanitized history of yoga is dangerous on many levels. If your yogic practice doesn't engage with the vast historical truths of yoga, if it doesn't make space for how these ancient practices were vilified—only to be co-opted by nations that perpetuated in quashing this knowledge so that Indians would lose their access *to be* well—then you are participating in an exploitative and unethical practice that is inherently, by definition, anti-yoga.

In early 2021, India was hit with some of the most devastating Covid numbers, and writer Meera Navlakha wrote in *gal-dem*, "On 22 April—the same day Archana Sharma was trying to find oxygen for her struggling family—India reported the highest daily increase worldwide (314,835) in Covid-19 cases since the beginning of the pandemic. Hours later, India's capital, Delhi, announced there were just 26 vacant ICU beds across the city. And, despite India being the largest Covid-19 vaccine manufacturer in the world, the country is facing a dire shortage of jabs as it is forced to fulfill export contracts to the US and Europe."[3] She added to these harrowing details: "All Indians have left are each other and their social networks. It is a damning reflection of the country's lack of preparedness to save its citizens."

This is a civilization that had answers to its own wellness inquiries only to have them stolen and profited off of, while it now as a colonized society suffers. It's not right. On top of this, so many of us, like my family and parents, are still in a process of grief absolutely rooted in the devastating impacts that colonization has had on migrants. A disconnection from the homeland has given so many of us a disconnection from self and therefore an understanding of who we are, which leads to both a philosophical and spiritual dilemma that continues to impact the well-being and psyche of my people.

Given the surge of Hindutva supremacy in India, a fascist-led movement, it is an urgent reminder that a holistic understanding of yoga—and therefore India—is immensely necessary. Yoga itself is a varied discipline compiled over thousands of years of thought that sometimes contradicts itself. Similar to the dynamics of racial segregation, caste was born out of Hinduism in India and is "one of the world's oldest social hierarchies, ordering society based on purity laws. A person is born, raised, and dies in the caste their family is assigned to, with doctrines of karma and dharma justifying extreme differences in qualities of life,"[4] writes my cofounder at Studio Ānanda (an online archive on wellness), Prinita Thevarajah. Given the growing instability across India, it's important to understand that the origins of these ancient practices are still steeped in present-day atrocities. "Brahmin priests and teachers hold the highest status, and lower caste communities (considered 'untouchables') are made up of Dalits and indigenous Adivasi communities."[5] This is a living reality for Indians in India, and it's important that we carry this understanding as we move toward global liberation.

Though caste was constitutionally abolished in 1950, "marginalization against lower caste communities has been historically consistent, with segregation, discrimination in opportunities, and a higher violence rate common against lower caste communities. Research estimates crimes against a Dalit person occur every 18 minutes, with 21 Dalit women being raped each week in an act of caste supremacy."[6] The last figures are the most jarring and go back to power and oppression used against femme bodies across the world.

"The right performance of the duties of any station in life, without attachment to results, leads us to the realization of the perfection of the soul,"[7] writes Vivekananda. This dislocation from the truth of spirituality can be seen throughout the world. Muslims are misusing the words of the Qur'an from Saudi Arabia onward. Buddhist monks are killing Rohingyas in Myanmar. I'm not trying to moralize—humans, by nature, are flawed. This is why we must engage with ourselves holistically so that we can continue to tell the real societal truths that haunt us but need to be faced. This means all of us get to talk and all of us are required to listen, *to each other*. No voices are prioritized over others, so we have a chance at healing by telling the whole truth, our whole truths. We deserve to return to the people we were before we were told we were inherently wrong. We need to return to a time when we were invested in ourselves as people; when it mattered if you were a good person. If that time in history has never happened, why aren't we all actively jumping toward that possibility? Is it just our unevolved and unchallenged nature holding us back? Or is it something more? Like our collective traumas?

◎ ◎ ◎

Yoga originates from practical and personal teachings documented in the Vedas, which are "considered the most sacred and treasured spiritual texts of India," but it is the handbook of the Upanishads (which contains over two hundred scriptures) where wisdom of the Vedas is transferred into personal and practical teachings. This practice was slowly refined and developed by the

Brahmans and Rishis (mystic seers), and the most renowned of the yogic scriptures is the *Bhagavad-Gîtâ,* composed around 500 BCE. "Though it virtually ignores postures and breath control," writes David Gordon White in "Yoga, Brief History of an Idea," "devoting a total of fewer than ten verses to these practices. They are far more concerned with the issue of human salvation, realized through the theory and practice of meditation."[8]

Over the years interpretations of yoga have shifted, evolved, and changed, but at the foundation, White suggests, "Yoga is an analysis of the dysfunctional nature of everyday perception and cognition, which lies at the root of suffering, the existential conundrum whose solution is the goal of Indian philosophy."[9] Even still, this history is complicated. White also adds, "The gulf between yoga practice and yogi practice never ceased to widen over the centuries. In later, esoteric traditions, however, the expansion of consciousness to a divine level was instantaneously triggered through the consumption of forbidden substances: semen, menstrual blood, feces, urine, human flesh, and the like." White explains the "menstrual or uterine blood, which was considered to be the most powerful among these forbidden substances, could be accessed through sexual relations with female tantric consorts. Variously called *yoginis, dakinis,* or *dutis,* these were ideally low-caste human women who were considered to be possessed by, or embodiments of, Tantric goddesses."[10] It's a complicated history that dips in and out of tradition and focus, which is why it's extremely interesting to track these shifts of perception that rely on caste in ritualized aspects as well.

Later "a new regimen of yoga called the 'yoga of forceful exertion' rapidly emerged as a comprehensive system in the tenth to eleventh century, *hatha* yoga is entirely innovative in its depiction of the yogic body as pneumatic, but also a hydraulic and a thermodynamic system."[11] This shows that as yogic principles shifted, discourse about evolution did as well. And yet, when the British rule began in 1773, according to sociologist and yoga practitioner Amara Miller of the Sociological Yogi, the *hatha* yogis "were associated with black magic, perverse sexuality (based in tantric philosophy), abject poverty, eccentric austerities, and disreputable, sometimes-violent behavior. The British government banned wandering yogis, trying to promote more 'acceptable' religious practices such as meditative Hinduism common among the educated and upper classes."[12] Meaning a quieter, more subsumed, version. These policies were supported, of course, by wealthier Indians who hoped assimilation would save them. "As the scope of colonial police powers grew in India, poor *hatha* yogis were increasingly demilitarized and forced to settle in urban areas where they often resorted to postural yogic showmanship and spectacle to earn money panhandling. From that point onward, "*hatha* yoga practices became associated with the homeless and poor, and were considered by both the British and Indians 'not only inferior but parasitic on other, worthier expressions of yoga that foregrounded meditative traditions.'"[13] It's interesting to see how *hatha* yoga has been propagated in the West, completely devoid of any of this history of its tricky origins and complicated evolution.

Yet we must note these changes and adaptations in order to hold the multiple truths of these regions and understand South Asia more complexly. Swami Vivekananda was a Bengali Hindu monk and chief disciple of the nineteenth-century Indian mystic Ramakrishna, who was also a Bengali Hindu mystic. From his teachings, Vivekananda learned that all living beings were an embodiment of the divine self; therefore, service to God could be rendered by service to humankind, a beautiful and integral sentiment.

"Our various yoga do not conflict with each other; each of them leads us to the same goal and makes us perfect; only each has to be strenuously practiced," writes Vivekananda.[14] I understand "strenuously practiced" as the diligence to continue evolving, to grow, to learn. And though there are many anti-casteist movements within Hinduism, it's particularly pertinent for those who teach and engage with yoga in the West to understand that white supremacy and caste supremacy are interlinked and that yoga carries its own legacy of violence. To deny the past is to actually deny the total reality. We can't keep living in these fractured interpretations of ourselves or our culture. Yoga has its own history that should be known by those who practice yoga. Much like our fascination with Greek or Roman mythology—how we interpret with reverence the cultures of Europe which details our attention is what I'm asking for. Yoga has evolved immensely and has had many different interpretations and schools of thought. It has depth that should be known.

I think how intriguing and beautiful it is that the very word *Upanishad* in Sanskrit is interpreted as "sitting down beside," *upa* meaning "near," *ni* as "down," and *shad* as "to sit." The translation bears an important remembrance. Jacques Derrida tells us that the word "archive" comes from the ancient Greek *arkheion,* meaning "the house of the ruler." So, the audacity of historical authorship is such that the empire sustains a narrative of necessity. Without explaining how it plundered wealth, depleting colonized lands of their resources, only to feed their own bellies. This was done through cultural and spiritual warfare—erasing a society's capacity to use its own knowledge is the domino slap of colonialism's reach. So, remembrance is important, as is correcting relationship and behavior.

In the Preclassical stage, yoga was an amalgam of various ideas, beliefs, and techniques that often conflicted and contradicted one another. The Classical period is defined by Patanjali's *Yoga Sūtras*, the first systematic presentation of yoga. When I lived in Montreal, I volunteered at a white-owned yoga studio. I can't remember encountering one South Asian person in a class in the years I worked there, which is something that bothers me most about modern yoga—how much this healing modality has been stigmatized even within my own disparate communities, adding to the self-hatred that so many South Asians I know contend with. That violent severing from our cultures, a legacy known as Partition that by and large still hasn't been faced by generations of people. One day, I was in a class with the owner of the studio, and she approached me and poked my belly. "You have a weak core," she told me, coldly. "I know," I said, embarrassed. That was it. She had no advice, no help—she decided it was completely normal to target the one South Asian (let alone only brown) person and humiliate them rather than give constructive feedback. I found this dismissiveness of my body an extension of whiteness, of the tonality of Bernier, who can only "categorize" other bodies. I would go on to cruelly punish my body for the next few years, obsessed with why I couldn't just strengthen my core. All these white women could do it—what was wrong with *my* body?

"I think about Ayurveda . . . how there are different constitutions and what suits one may not suit another . . . but that too is part of our Indigenous wisdom, a variety of practices for all the different needs," Lakshmi Nair said in an interview with Studio Ānanda.* "What is colonizing about the way Western yoga is usually taught is that it treats all human bodies as if they were the same or based on some idealized human body . . . I think the Indigenous approach respects the vast diversity of creation while maintaining a grounding in the underlying Unity. The colonized viewpoint is kind of the opposite."[15]

This is perhaps the most depressing part of the whitefication of yoga. White bodies suddenly become the norm, while an ancient practice that was about purification of the soul to attain a sense of enlightenment and closeness to the divine has now been co-opted to become a physical practice about attaining a "hot bod." The way yoga has been stolen is not a unique example of the way money and capital blind us from our social responsibility, but it's one of the most glaringly obvious cases. A couple of years ago, a white woman wrote a jeremiad of watching a bigger Black woman do yoga in an overwhelmingly white classroom. It felt violent, paternalistic, and fetishistic to not only speak *of* someone while you watch them under the guise of care, but then to write about it so unabashedly to gain sympathy for your unconscious whiteness. As if yoga was for white women only. This claim that whiteness has enforced, the claim all white people seem to have to *the other,* always centering themselves in every narrative, is a very big problem that needs to be addressed and unlearned. This has to be included in the necessary evolution of yoga.

* Studio Ānanda is an archive of wellness, as well as a social practice, co-founded by Prinita Thevarajah and me.

If white people spent more time humanizing themselves, maybe they could see non-white people with more depth and complexity, as well. This means not treating us like we are the other, unconsciously objectifying our difference, only to expect sympathy for the very definitions that were created *by* white people. All Black, Indigenous, and other people of color deserve your respect and humanity, not your *sympathy*.

If you are a white person who profits off yoga, what we need now is your action, your commitment. Find trusted organizations and give at least a quarter of what you make to support on-the-ground work in India, either with Indian farmers or non-casteist communities that are working to protect land and water and culture. Make an effort to support organizations that are safeguarding farming rights and practices to be with the land, to help maintain its biodiversity. You must reinvest profit in the communities who have been robbed of it. There are always ways to give back, to create holistic structures of care that are oppositional to methods of extraction and depletion.

Things can change, but it takes a collective effort to actually shift society. "I imagine if we acknowledged that everything we consume is the gift of Mother Earth, we would take better care of what we are given. Mistreating a gift has emotional and ethical gravity as well as ecological resonance," Robin Wall Kimmerer writes in *The Serviceberry*. It's not about *not* using these modalities but, *rather*, about honoring where they come from, and creating a healed and equal relationship with them. "The currency in a gift economy is relationship, which is expressed as gratitude, as interdependence and the ongoing cycles of reciprocity," Kimmerer writes. "A gift economy nurtures the community bonds which enhance mutual well-being; the economic unit is 'we' rather than 'I,' as all flourishing is mutual. . . . Those who have to give to those who don't, so that everyone in the system has what they need. It is not regulated from above, but derives from a collective sense of equity and accountability in response to the gifts of the Earth."[16]

This, I believe, is how Indians have viewed yoga. It was to share, to give, to teach. Of course, within that there were structures of power, but we have the ability to create yogic systems that are for all. That take into consideration trauma, such as rape and sexual abuse; or the consideration of community, investing in healing the community, not merely profiting from it. If we want to learn how to create good karma in the margins of colonization, outside of casteist rhetoric, at the very least we must start telling the truth about all of it.

So, are you being honest? What would it mean for us to live more honestly and in communion with each other?

Charles Eisenstein writes, "Gifts cement the mystical realization of participation in something greater than oneself which, yet, is not separate from oneself. The axioms of rational self-interest change because the self has expanded to include something of the other."[17] How do we become a culture that builds horizontally, with each other fully in mind? I guess we just start. Yoga is a good place to start to understand and believe in the duty we have to each other. The duty we have to our own collective evolutionary dance and motion.

CHAPTER 8

ON IBS

My gut issues started at fourteen. The blistering palpitations, the wretched unruly heat my stomach would generate like a nuclear bomb testing site. It would take over my life when it came, penetrating every pore, causing so much chaos in my body, and pain that would last for days on end. After the first few months of this, I began to see and feel this visceral connection between my mother and me whenever I had an onset of IBS, like she could short-circuit my body, bringing me to my knees, just through her mind. I could feel her even when she wasn't close because that's how connected we were. That was the first explanation I had for my IBS, even as a teenager. It didn't make sense, but in my universe, I guess it did.

When I first spoke to my sister about what was happening, I would tell her it was like I could feel my mother's rage in my stomach, like a galactic swirl except it was a black hole of cramps. Like a bat signal in my body, it would come and go, at a whim, exiting me in shreds like a hurricane leaving a small town.

The "weak core" my white yoga teacher pointed out is a result of trauma—intergenerational and my own—which is a stored physical energy in my body that was perhaps created as an energetic entry point by my mother. I don't know how else to explain it other than there were no boundaries between us. I was so porous to her, and because she controlled every action and watched every move, my body became controlled by her moods and needs. My IBS, I believe, became a direct embodiment of this. I felt so powerless around her, primarily because I had been conditioned to believe, by her, that she was an authority figure, a god. So much so that she could punish me even when I wasn't physically close to her. I was the perfect, compliant captive because I believed her. I believed in her power over me. In the chakra system, IBS is directly located where the solar plexus resides, and a blocked solar plexus represents low self-esteem, shame, and unworthiness.

I mean, you can't make this shit up.

In the worlds of health and wellness, there's an inconsistent belief that if you try hard enough at something, eventually you will see results. What's so dangerous about this thinking is that when a body can't do certain things, because of its sometimes (un)known limitations, that body is then (subtly or not so subtly) societally ostracized. Just by virtue of the world being catered to able-bodiedness, we forget bodies are still valuable irrespective of what they can or cannot do. Chronic pain became a safe way to engage with my body at a distance. It was an added buffer, a shield for me. When I was focusing on one pain, I wasn't looking at the other. As the Buddha reminds us, "If the body is not cultivated, the mind cannot be cultivated. If the body is cultivated then the mind can be cultivated." I wasn't finding balance because I was in constant avoidance of myself—the body's pain was a way to deflect the mind's trauma.

Yet the irony is that your body keeps the score because your mind wants to forget the tally. Even if we are unconscious to it, what we experience becomes stored quantitative data. As a child, well into adulthood, I lived in fear, all the time; I never felt safe. I became hypervigilant to my body's sensations—yet I was also completely confused by what I felt, the pain didn't add up to the memory, or vice versa. Such sharp attention and awareness are a product of abuse, as your body is always tuning in to the possibility of attack, but because nothing has identified the root of the issue, you go into overdrive, believing everything could be a possible violator. I had to be prepared to defend myself at any time, whether physically, emotionally, or mentally. Living like this was an impossible reality. The more I thought about it, however, the more I came to realize that the gut is the home of intuition, and therefore I had to learn how to trust mine.

Now I understand that my mother's abuse had created a bond that had morphed into a portal that linked me directly to her. I hated that she could do this, I hated how thin-skinned I was. I already felt surveilled and on edge constantly, frightened that she was always watching me. All of this resulted in a hazing of my body, especially where my stomach was, and all the sexual parts of myself. If I veiled myself, if I self-actualized into a blob, something out of Hayao Miyazaki's universe, I wouldn't have to feel her conquer me. When I started cutting, it became another distraction. If I cut myself somewhere I could control, then the pain would numb the stomach pain. It was a constant cyclical void of pain I existed in. Instead of listening to what my body was trying to tell me, and not knowing I could learn how to listen to her, I turned on myself.

This became a dynamic where I was then mistrustful of myself. Because the pain in my insides was not being validated by an authority figure, I was forced to believe that I was fine. When I would say I wasn't, my mother would eye me suspiciously, cocking her head, only to tell me all the places she was hurt, but did I see her complaining? My father would deflect, kindly telling me I'd feel better soon. And why wouldn't I be? I wasn't starving. I had a home, two parents, food on the table. I knew I should be grateful, so I was. "The good girl believes something like, 'One day Mommy/ Daddy will be full and then they will give me what I need,'" Bethany Webster writes aptly in *Discovering the Inner Mother*.[1] The problem is that day sometimes never comes. Instead we become "adults who experience high levels of stress when faced with everyday situations like disappointing someone, receiving compliments, setting boundaries and taking care of themselves."[2] So, then, some part of you, the good girl, begins to believe, because you are constantly told that your parents know best, especially more than you, that one day they will see you for who you are. See how broken and frail you are (and yet how hard you try) but how much you really need them. How all you want is their good, loving, safe touch. You want to be a child.

As an adult, this will mean that people won't even see the strain in your body, and like your parents, all they'll see is what they deem as success, never once questioning the impact it has on your mental, physical, emotional, spiritual, and sexual well-being to carry this much. To never be able to complain about what the cost has been on your psyche. So, as a coping mechanism, because your parents don't witness you all the time, something inside of you begins to disconnect from your pain, your sense of reality has to mimic theirs for your own survival. You begin to tell yourself that your needs are too much, and you convince yourself that the more perfect you are—that the more wholesome, morally upright, intelligent *you are*—you will be loved, cherished, and adored. So, you go on a hamster wheel, trying to perfect yourself. In that process, you forget you are unwell, and that you have needs, too. And you keep forgetting, and keep forgetting. I've taught myself to perform in the face of extreme pain, dulling my senses. I've taught myself how to numb my body's quakes and shivers. I've hidden my true self, denied my body its right to speak *to me* and participated in my own silencing again and again and again. Until, very recently, I realized, it didn't have to be this way.

I took my IBS as a sign of my inherent weakness. As if I was too broken to be given a real body, one that wasn't always in revolt. The ableist ideas of being healthy were drilled into my brain (as well as my sister's), and the more our bodies revolted, in different ways, the more we continued to be silenced by both our parents. Or made fun of, as when our food allergies, bodily reactions, were always read as exaggerations, instead of real things that had been determined by our own bodies because we knew them best. My parents would call both my sister and me "extremists" because of our sensitivities to gluten, sugar, and dairy. So, I was forced to continue to eat "normally" or "moderately" so that I could grow accustomed to it, always being told that what my body knew to be true was again, inherently, wrong. Or at the very least, amenable.

I wish I had known earlier that if you compromise your own unknown needs for the tenacity of another's, there is always a personal cost. So, the more I was *that* Fariha, the one that was together, composed, kind, but hidden, obstructed even from myself—the less I could be the full Fariha, the whole Fariha, the one that was also tender, angry, frustrated, tired, but desperately, sickeningly, wanting of love. According to Webster, "The 'used child' within us longs to be loved for her real self, not just when she is wearing the mask of the 'good girl' and demonstrating patriarchal values (productive, perfect, conforming to expectations, making others look good, sacrificing, suppressing, etc.). The used child longs to be loved when she disappoints you, when she is grumpy, when she is inconvenient, when she is messy, when she is confused, when she produces nothing, when she is inconsistent, when she is empty-handed, when she changes her mind."[3]

The reality—I also found—is that most adult humans I've encountered aren't very secure. Projections are always being placed. I found that I kept getting into circumstances where I would be intentionally misunderstood, or that I'd be put into positions where I would have to apologize for things I hadn't done—just to keep the peace. This was my role as the negotiator; I was always the person who compromised, extended a hand, even when I was usually the one being misunderstood. This caused me to further shut down, to get deeper into the performance, out of a sense of defeat that no matter what, I was the one who would have to succumb to other people's power over me.

In my life healing has been about integrating the fragments of mind, body, and spirit that were shattered by trauma. As I nurtured and braided all these disparate parts of myself together, a more honest version of me began to form. The more I thought about this on a micro level as an individual, the more I began to see a parallel between the Earth and humans. If I could heal myself—and what I mean by this is largely about acceptance, it's healing to accept a failing body and tell that body *I love you, unconditionally*—maybe we would have more proof for a way to collectively heal the Earth, too. Because the Earth, like our own bodies, needs so much love, nurturance, and gentleness from us.

◎ ◎ ◎

We are far more connected than we seem to think we are. To our families, to our ancestors, to our land. The path to healing as a South Asian is a multilayered process. To deny us of our cultural roots, meaning our cultural and spiritual customs, means to rob us again and again. To keep us distracted, forgotten, is to further the poison. To then co-opt, steal, and appropriate our culture is an ugly, unequal, and devastating consequence of a malice so great that to quantify it would mean an endless, Nuremberg-style trial of naming the poison of the atrocities done to us. Repetition is the mute language of the abused or colonized person. The reality of it is that to move through this, to heal, societally, to come to ourselves, then to heal our bodies, means healing our lands. Something that all of us, especially if you are a white-bodied person, must contribute to. It's your karma and it is your responsibility to help *us* heal—that is the most significant gateway to your own healing.

"Displacing is a way of surviving. It is an impossible, truthful story of living in-between regimens of truth,"[4] says writer and filmmaker Trinh T. Minh-ha. So much of this book is about filling in these histories for other South Asians, so that we may claim what is ours, and instill a cultural recognition that we are better when we work together, when we can see the rich majesty of our lands, of who we are. We can't afford to keep forgetting—it's time that we remember. Being South Asian has sometimes felt like making work in exile. A lot of us have been taxed to reimagine what our futures are, without knowing exactly where we come from. We are constantly faced with the gaps in our memories, because of our incapacity to face ourselves. But I think it's time to look at these wounds. "Awareness is not just in the mind, but also includes body knowledge,"[5] Gloria Anzaldúa writes in *Light in the Dark*. Grasping my past, my ancestors and their past, collecting stories through my father, has been a way to understand myself, to root myself into something. Through understanding them, I've begun to understand myself and why my body has been revolting for so long. My IBS was a way to understand my body on its own terms, not what I, or society, determined for it.

My parents' attention was both harmful (from my mother) and negligent/nonexistent (from both). In *Care Work,* Samarasinha reminds us that most of us "had received shitty care, abusive care, care with strings attached."[6] And most of us have denied situations of vulnerability, opting for strength as the constant mirage. I imagine a lot of people look at me and see some kind of vulnerability and assume the ease that comes with it. It used to spill out of me like toxic waste, and I would have to embarrassingly clean up the aftermath, vomit on the floor. The number of times my vulnerability has been outright rejected, castigated, humiliated, is endless. In New York, a lot of shitty encounters with people made me think that I had to harden myself. I've seen how easily people will blame me for things that have nothing to do with me, and it happens *often*. How quickly I've been relegated to "crazy" by former friends and lovers. That is the pursuit of an authentic life. To become yourself, totally, means you lose a lot of people along the way, because most people are incapable of sitting deeply with themselves. I've had to come to terms with that. I wasn't the problem. Now, I've realized boundaries are not only necessary, but that not everyone even deserves all of me. I don't have to be liked, I don't even have to try to be liked. I'm enough, just as I am.

When you're a child of abuse, of trauma, what no one tells you is that you have to look out for yourself. That in itself is a product of the abuse—you are neglected. Let alone your body. You have to learn what a checkbook is, how to do taxes on your own, how to call 911 to save your own life. No one prepares you for the world, holds your hand or invites you to Sunday family dinners, celebrations, anniversaries. Birthdays become a tenuous, heartbreaking thing, and no matter how many people tell you that they love you, there will still be those who won't understand what it feels like to be in a constant void, to exist in lovelessness because the core of you yearns for that love you will never get. A mother's cognizant and lucid love. That's what it's like to always feel alone. It's to wonder who would save you if you were in the gutter right now and not have immediate, tangible answers.

Recently a healer told me that I shouldn't hold on to the pain of my life and childhood. I initially agreed, but then afterward felt a rage brewing. What people don't seem to understand when they tell me this is that ever since I showed sadness at ten, everyone has wanted to tell me where to put it. Their discomfort with seeing a sad child usurped their desire to actually tend to the child's needs, to question how I was feeling. I've been abandoned by more people than I haven't. That hurts extra when you don't have a stable family unit, when you don't have a mother to cry to, turn to, learn from. This, not so ironically, is all related to the solar plexus. Feelings of betrayal, problems with forgiveness and a lack of safety in oneself, a difficulty with staying in one's own personal power. Recently, I realized that my IBS had given me the infrastructure to unlock a lot of answers.

The more I've turned toward myself over the last few years, the more I become and reenvision myself as my own being. Learning to understand that my IBS was trying to speak to me, that it was reminding me that the PTSD and panic attacks weren't a reflection of my hysteria—but rather a communication that I needed to decipher—made me realize the vast layers of knowledge that are coded in un-wellness, in chronic illness. What if we listened? What if we paid attention? What if that's all that healing is—just listening to yourself? If so, what are you trying to tell yourself?

So much of my life I lived with this narrative that my body had failed me, but I mainly felt this way because I had been fed enough crap about the imperfection of my body since I was a child that at a certain point those criticisms adapted into my own thoughts. My mother would castigate me, calling me *dhumshi,* which is analogous to "fatty," something my sister would also call me. Imagine a society that didn't feel the need to critique everyone's outward appearance, imagine if we cared less about judging others and focused more on how we can be active participants toward bettering our collective humanity . . . imagine. These days I rest on this hope, because I take solace in my own reckoning, in my own leaps of faith and consequence.

I keep thinking back to when I was eighteen, a few days before my nineteenth birthday, realizing I was pregnant. I had known for a while, or at least suspected it, and that was the next stage of disassociation from my body. I didn't understand why it couldn't be normal like other girls. Why was I always bloated? Always uncomfortable in my skin? I didn't want to live like this. I wanted to learn how to accept myself, I wanted to love what my body *was,* not what it wanted to be. Which was a journey in and of itself, because my body, itself, has potentially always been in flux, and maybe that was OK?

The IBS, as it has stretched out for almost two decades of my life, has recently become an exemplary signal that I follow. This chapter, in fact, has been the most difficult one. As I write these pages, I'm feeling them out, too. It's been over a month, close to six weeks, since I've had an IBS flare-up this time. It's made it hard to digest (quite literally and figuratively), but the lessons that are revealing themselves are immense. It feels strange to admit this, but my body, these days, feels like a blessing—completely my own in her imperfection, completely my own in her complexity. The lessons she gives me are designed for me. How lucky that we all get the machinery to liberate us from our ancestral patterns. Your body is speaking to you. I just had to learn to understand my own language first.

◎◎◎

There's a humility for me in unfolding the mystery of my body that reveals itself a little bit more each day. The chronicling, the tracking, and the achievement of listening to myself with openness and care has been profoundly healing. Because, with this, I've also accepted myself. It's a continuous journey of not assuming that I know an answer when I haven't even checked in or asked my body first.

Just before the onset of this latest bout of IBS, I began to compost again after my dear friend Zenat advised me to reintroduce it back into my life. As a child who was raised with composting, I felt sad that I had forgotten my ways. I felt sad for my mother, and I realized one way I could honor her was through remembering her for the good, valuable things she taught me. Since I was a child, she's been composting, gardening, tending to Mother Earth's body, but not her own. Maybe it's easier, safer, to do so when you're unwilling to prune the gardens of yourself. Not surprisingly, my own ritual of composting began to shift something inside of me, as did paying attention to the cycle of the plants in my apartment or the freshness of fruits and vegetables in my fridge. The more I attuned myself to my environment, as well as the other living beings in my environment, or even the dying plants that I was now no longer facelessly discarding, something inside of me also became more attuned to my patterns, and thereby the pattern of life. The cycle that connects us to the rhythm of the wind, the quiet purr of fire, the ways seasons change, and the surf of the ocean's whip.

There's a softness and agility I've adopted through the cycle of composting—it has an intentionality to it, one that I crave, and a remembrance attached to it. Finding small ways to turn inward, every action, even a small one, can become a ceremony. I'm learning that there are so many ways to honor the physical, and a part of it is honoring its voice. I'm learning that the Ayurvedic diet is my people's diet, and that ghee and kichari (two things that have helped my IBS) are both integral parts of my culture's healing that I'm now accepting as my own. The ways of my people are bringing me back to myself. Balancing ecosystems, whether on the land or in the gut, was an integral part of the cosmologies that grounded our ancestors and an important reminder we need now. Disconnection from the land is disconnection from our gut. It's important to know how your body responds to the food you feed it so you can be cognizant to yourself, and the world, holistically. I'm remembering this, through the act of healing I'm reclaiming my DNA. I'm remembering my people had it figured out. After a millennium of civilizations quashed under colonial tyranny, it's important we remember what we were forced to lose.

So, I'm turning to them, my ancestors, I am learning to prioritize the ancient knowledge that determines the importance of the sacral center and the wisdom of my gut. These days, I remind myself that all I can do is be present with the situation at hand—to trust the cycles. I do as Bri. Maya Tiwari, Ayurvedic teacher and writer, did—I make a silent vow never again to sever my ties with the Mother's rhythms. I commit to myself, to this body, and this Earth.

PART III

ON SELF-CARE

The most anti-capitalist protest is to care for another and to care for yourself. To take on the historically feminized and therefore invisible practice of nursing, nurturing, caring. To take seriously each other's vulnerability and fragility and precarity, and to support it, honor it, empower it. To protect each other, to enact and practice community. A radical kinship, an interdependent sociality, a politics of care.

—JOHANNA HEDVA, *SICK WOMAN'S THEORY*

CHAPTER 9

INTRODUCTION TO RADICAL SELF-CARE

> One of the best guides to how to be self-loving is to give ourselves the love we are often dreaming about receiving from others.
>
> —*bell hooks*

Though I've been writing about self-care professionally for about seven years now—diligently, methodically, tending to it like a spiritual study—at times I have felt I know nothing about how to *truly* care for myself. Initially, I turned to the concept of self-care for answers, and found the works of the late, great bell hooks and the tender radicality of Audre Lorde. I, too, craved an anecdote to better understand who I was, and I longed to know how to aid myself in order to become that person. I knew that the reservoir of unresolved anger rooted in me like a deep prickly weed made it hard to know how *to love myself,* and so, I assumed (rightly) the first step toward my higher, evolved state was to learn myself intimately and to accept it all. I had to understand myself like a lover and appreciate everything I felt was unlovable, gaining a security I had never known in myself, and according to hooks and Lorde there is an inherent radicality to caring for yourself when you come from a lineage of oppressed peoples. Taking on self-care as an active embrace has meant merging the needs of my mind and body, because in the act of self-care—the mind and body are prioritized, that is the very self you are caring for.

Nothing that I share is a one-size-fits-all theory, but instead something I've gleaned from my personal studies. I have often felt like self-care should come with instructions, because I didn't quite know where to begin the process myself. The nature of self-care's commodification has meant that we've lost track of how personal this journey is. We all have our own traumas, fears, needs, and therefore our own specifications, idiosyncrasies. With such an overwhelm of choice, it can be difficult to know what we need as individuals, how to *care for our own damn self.* Capitalism has destroyed our sensors, and instead we want it all—or much of it—without understanding what is inherent or honest to ourselves.

We take direction from websites, peer reviews, and best ofs to determine the scope of what we like. This is of course those of us who have the luxury to afford and fathom caring for ourselves. Why does wealth—real or imagined, inherited or self-made—make us believe in our own entitlement? In the 2015 *New York Times* article "The Price of Nails," Sarah Maslin Nir spoke to the labor conditions of nail salon workers across the boroughs of New York. "Nail salons are governed by their own rituals and mores, a hidden world behind the glass exteriors and cute corner shops. In it, a rigid racial and ethnic caste system reigns in modern-day New York City, dictating not only pay but also how workers are treated."[1] This is how they're treated internally, then there's dealing with actual customers. I remember feeling immensely grateful for this piece when it was published, as I had witnessed with my own eyes enough women (always white women) bullying nail salon workers in a variety of different scenarios to realize that there was a context of entitlement of who gets to self-care that wasn't being acknowledged. So much of the rallying cry of white supremacy happens in these moments of ubiquity, when even the most virulent acts of entitlement are ignored because we expect it of whiteness.

I can't write about self-care without first pointing toward the obvious—your care cannot impede on the care of others. Just like the concept of freedom—ask yourself, is it really freedom if it is only for some? If we prioritized not just what we think we deserve, but also how we are in relation to others, and thus how we care *for them* as well, we might experience true liberation. Being unbound by fear or hurt or pain, simply by showing up as the best version of who we are, sounds like liberatory behavior. Why has whiteness made the playing field so dirty with such high stakes and yet such low standards? Why isn't the measure of a successful society how well we care for each other? And how can we possibly believe that programming people to think only for themselves could result in holistically positive results?

I think all these years I struggled with true, unadulterated self-care because I was taking direction from others, expecting them to have solutions that I could've easily just learned by myself, for myself. But I believed that I would be cured by another's expertise. When that failed me too many times, I began to realize that I could find the right acupuncturist *for me*, the therapist *for me*, and that I could build my own routine around my life and my needs. Agency is an important and necessary part of self-care, because in the process of learning that you have it, you are forced to take ownership over your life. Caring for yourself means taking a giant leap toward yourself. You have to put yourself in your own driver's seat.

I did not know what "life force" was until my first Ayurvedic doctor, Pratima Raichur—an absolute renegade of Ayurveda in America—introduced the term to me, explaining it through the use of the Sanskrit word *prana,* which means "life force," to explain ownership of oneself. Mine, she explained, was low and dull, which was evident in the hue of my skin's tonality (I was yellow, but if I were more balanced, I'd be red) and the dryness of my body and hair. Details, like eczema, that were dismissed by medical experts were to her important clues about my constitution. My dryness was an indicator of my incapacity to absorb nutrients, even good ones. An apt metaphor, it suggested I had an excess of fire, meaning I had to focus on cooling my body, which meant doing less vigorous exercise and embracing an overall slowness, including reducing caffeine intake. As she told me this, I protested, almost resentful. "I can't do that," I told her point blank. Her advice, like a lost grandma's ardent wisdom, was blunt: If I wanted to truly heal, she told me, I'd have to start registering that my body had answers. The markings were telling her what she needed to know.

Prana, she told me, is also about will.

Will is an interesting concept, because clearly I have it, I escaped my life and built what I have on my own, and yet on a fundamental, day-to-day level I felt so broken, so unchosen, that the IBS ravaged. Everything was a representation of my mother's lovelessness and abuse, and any relationship could become a rearticulation of it.

◎ ◎ ◎

In Chinese medicine, *jing* is known as essence—another interpretation might be *prana*—and it is passed from parent to child. Prenatal *jing* nourishes the fetus during pregnancy and determines basic constitution, strength, and vitality and is stored in the kidneys. One way to know if you have a weak *jing* is to examine and know the state of your kidneys. The last time I was alerted to take notice of my life force was during an ayahuasca ceremony, just before the pandemic. At the time, a teacher I was working with explained that my life force had been hijacked at such a rudimentary stage that to think that I could be in control was a new paradigm for me. But we all perform safety, don't we? Especially in ourselves. We think we understand who we are, and yet everything we know has been informed by others on some foundational level, in ways we can't even comprehend. I had to notice that I had tied my own hands (or, rather, that I had the keys to my own escape) simply by acknowledging the truths that I had never uttered, facing the reality of how I participated in my own life's rhetoric. I had to accept that most people didn't really understand me because they couldn't see the psychological and emotional ravaging, but that didn't mean it wasn't there. I needed to learn how to validate it to myself.

Living in the Western world gives you a certain hubris that you have choice, or freedom, and that things aren't silently forced onto you, chosen for you, spoken for you. Whether it's the laws that regulate you, the state that surveils you, officers that police you, or capitalism that seduces you, at the end of the day, you are bought and you have to be conscious about so many different facets of your life to understand how and when you are bought. In order to not fall prey to losing your sovereignty, you must choose the life you want, outside of the confines of what you've been told *to want.* Tricia

Hersey, founder of the Nap Ministry,* explained in an interview with *Atmos* magazine, "Our lack of resting has stolen our imagination and our ability to be inventive and subversive and to imagine and to have hope. And to me, that's true oppression. To me, once you have taken away a person's ability to see their way out of a situation, to see a new way, to imagine a new world, to see something different, to invent, you pretty much have them. And I think we're at that right now."[2]

This year has been a year of rest for me. Of borrowing rest from the years I had thrown that need to the side, not believing that I could be a successful writer (as in, be comfortable) and not need to worry for the next day, week, month. In my mind, I didn't feel like I deserved to rest, because I hadn't received the monetary foundation of being afforded the time *to relax*. Of course, this is a trauma response. For so long I just thought it was my drive, or the plight of being a Capricorn stellium (listen, it's hard out here), but it was really just my incapacity to stop working out of fear that I would lose everything faster than I could say anything to protest.

We are road maps to ourselves and what *we need* on an individual level. In a recent ayahuasca ceremony, I had been ruminating on gentleness without completely understanding what that meant. I knew I longed for a soft touch, an agile hand on the head, fingers

* "The Nap Ministry was founded in 2016 by Tricia Hersey and is an organization that examines the liberating power of naps. Our 'REST IS RESISTANCE' framework and practice engages with the power of performance art, site-specific installations, and community organizing to install sacred and safe spaces for the community to rest together. We facilitate immersive workshops and curate performance art that examines rest as a radical tool for community healing. We believe rest is a form of resistance and name sleep deprivation as a racial and social justice issue." "About," *The Nap Ministry*, accessed March 25, 2021, https://thenapministry.wordpress.com/about/.

running through my hair with ease, but as I was unpartnered and celibate, where would I find this intimacy? Grandmother, the medicine, revealed that it had to begin with me. I asked her how. She pointed out that I was lying on a hardwood floor, rolled over in pain, convinced I was incapable of doing anything about it. At that moment, my life force was weak. She pointed out that there was a mattress, a perfectly good clean mattress right behind me that I could ask for help to get onto. Sometimes the medicine makes it hard to do the most benign things, so I hesitated. I would have to ask for help, and that scared me. It took me a while to ask, to even muster the word *help*! The first time nobody heard me, and I was immediately embarrassed, silenced, my abandonment wound perking up. But Grandmother kept pestering me to ask again and again—she told me to ask however many times I needed to until I was heard. So, I did. Finally someone heard me and eventually helped me get onto the mattress. It was an extraordinary feat, and as I lay there, on the mattress, my ass no longer in pain, I saw my own pattern playing out. Then I realized another way I obstructed my *prana* was not believing that I deserved help or deserved to ask for things.

No longer tender from the hardness of the floor, I cried at how easy it could be if I learned how to ask for help. It was, truthfully, something I needed on such a crucial level of self, and the absence and depletion of that necessary tenderness, the tenderness to believe you deserved help, had made me parched for love, for affection and care. So, I remained in a state of desperate craze, perpetuating my own myth. This is what happens when you are taught that your value is how well you accommodate others. I thought of the words of CJ Hauser's "The Crane Wife" essay: "When a woman needs, she is *needy*."[3] To want gentleness, Grandmother told me, was key to the next step of my evolution. I had to believe not only that I was worthy of the love I wanted, but that it was also possible to find the people who could meet me here. But it had to start with me. More than anything, I had to give and offer this love to myself.

Regaining my life force was my first actualized step of self-care.

Earlier this year I remembered my father taking my mother, my sister, and me on a drive. He entered the woods—I don't remember how it happened, but it happened fast. I think this was around the time he was first rebelling, too, thinking that he could control her hysteria if he wielded some kind of power as well. I don't know, maybe that's my read on it all these years later. At a certain point, past the clearing of trees, he started reversing toward a body of water. It happened so quickly. It was as if he knew the bank intimately and had been considering death here, backing into the water as my mother screamed. For most of my life, death at the hands of a parent felt so close, so immediate. When I think of life force, and where I lost mine, I think of all the times in my childhood I witnessed extreme violence that then suspended me into a dis-reality, a chaotic amalgam. Incapable of moving, of saying no, of screaming. As a child who relied on being "good," my silence, my self-numbing, caused me to lose my life force. Every time I didn't scream, every time I didn't defend myself, every time I let my body get hit, I was allowing someone to determine my path for me. I was giving away my essence of self. This is why, I believe, survivors of rape and sexual abuse have a hard time expressing what is lost. What you lose is part of yourself. Sometimes you give so many parts away, however, that you're no longer accessible to your own self.

When I began therapy, I had to be reminded over and over again that I had ownership over my life. I didn't realize that I was living in perpetual fear, that my anxiety about something going wrong, about someone hurting me, accusing me of something I hadn't done, was a form of PTSD. It took me a long time to realize a constant slow panic was a sign of distress and therefore a sign of something deeper at play. When I was a child, I just presumed I was an abomination. In *Dreaming the Dark,* Starhawk writes, "Psychologists have constructed a myth—that somewhere there exists some state of health which is the norm, meaning that most people presumably are in that state, and those who are anxious, depressed, neurotic, distressed, or generally unhappy are deviant."[4] That's exactly it—I presumed I was deviant.

Finally understanding that I had to give words to all these things and let them sit in my system, giving my cells the lifeblood they had been starved of for so long, allowing myself to claim the space that I had never been given, was a difficult point to arrive at, and I wonder if it's a place of constant arrival. Every day I gain further clarity of myself and my body's alchemy—all its mysteries and wonders—understanding that healing or wellness is not a stagnant state. For some of us with bodies in revolt, it is a state of unraveling that's necessary for the rebraiding to occur and reoccur. Life is an upward motion dance, and along with that, I've realized, wellness is, too. I used to get so upset by the slowness of my path, at how difficult it is to be in my body after all these years of trying. The work is glacial; it's punishing, too. The only thing that has taught me any calm is to surrender. To embrace all of it and let it unfold as it must.

"If you are struggling to practice self-care," disability rights activist Mia Mingus writes in "The Four Parts of Accountability," "you will inevitably have to confront why you consistently put yourself last. In the example of self-care, you may need to let go of some things or say 'no' to something or someone; you may need to assess how you spend your time and why you make space for things that are not what you ultimately want to prioritize. Transforming your behavior is hard work and is easier done with support. Find people in your life with whom you can talk about your accountability, mistakes, things you're ashamed of or feel guilty about, things you need to apologize for, or times when you weren't your best self."[5]

◎ ◎ ◎

Something about caring for myself has also felt erotic. I mean erotic in an energetic sense, like how Vedic priests and Sufi mystics write about tantra, the art of energy, harnessing sexual pleasure to exist beyond sensuality and tactile pleasure, but rather about tapping into something akin to the divine, a source for all, a place of majesty and mythic imagination. To me, real sensuality was inexplicable, or what Derek Walcott calls a "green thicket of oblivion," the energy that is both inside and outside of us—that's the one to channel.

Growing up in a culture completely devoid of eroticism—yet having such rampant sexual abuse—says a lot about what exists, like a penetrating disease, at the heart of South Asian communities. Sex can't be talked about, but it's everywhere it shouldn't be. On your mother's tongue, in your uncle's gaze, in your cousin's touch. So much of these worlds are hidden, but growing up I felt it everywhere, all the time. When my mother and sister told me to go up on the roof with my creepy uncle, I said no, firmly. In retrospect, it's interesting to see the times I had a backbone, but it's usually when there was a clear delineation between good and bad. With incest or rape, or all the things we pretend don't happen in our culture, sometimes it's murky if you know the person and feel you must trust them. Or if some part of you felt good to be desired. These are complicated feelings.

As I grew older, what I realized was that this attitude wasn't purely a product of South Asian culture. The literature, art, poetry, song, dance, even faith of these regions—and I'm thinking of India, a greater India, where we were all one—were erotic; this was once a powerhouse of eroticism, where thought about sex and desire was an evocative and ever-present current in the culture. What I had experienced in my youth was a mutation of my culture, of all the things that sex becomes when it is perverted. In Mughal paintings depicting homosexual sex acts, tantra, and religious sex acts, the legends of Jahanara Begum the lesbian, of *kali-sutra,* there is immense evidence that my familial lands pulsed with energy, with life, with *prana,* with sexuality. And now? What had happened to my lands and my people? Why had we moved so far from our cultural truth, where men freely hold hands on the street, are tender with each other? Where sexuality is an ingrained part of culture and space?

Part of reckoning what is lost inside of me—this ethereal lostness—is also tied to the immense loss of tantra, of sex, of our life force. It's existential, but maybe existentialism is just an inherited feeling. A feeling of loss so great that for generations people have remained, living in these dimmed versions of themselves.

Impacts of colonization—directly tied to the very puritanical Christian ideal of rigidity—colluded in Indian culture, inserting codes such as Section 377, a British colonial penal code that criminalized all sexual acts "against the order of nature," which prosecuted people engaging in oral and anal sex as well as homosexual activity. It has also been used to criminalize third gender people across the Commonwealth. "Sexual hierarchies, we found," writes Silvia Federici in *Caliban and the Witch,* "are always at the service of a project of domination." When you force people to feel morally corrupt, their desires ungodly, using words to demarcate non-white people's inherent bestial nature, while crediting the white race and the Christian faith (brought to us by a Palestinian Jew, lest we forget) with purity, you make a people turn from themselves even more. So many of us are wandering around, thinking: Where do we even begin to start to heal these wounds? For so long we've had to pretend the unthinkable didn't happen, that our lands weren't completely stripped by an ungodly and cruel force, and that does something to desire. It forces desire to become about power. When you feel you've been exploited, how else do you determine your worthiness than by stealing it in the same ways your captors continue to?

Radical self-care, I believe, means going to the beginning. I want to explain to you how I see these things, holistically, complexly. Naturally, when you feel you don't have agency, you are pliable to a market, to capitalism, to an industry that will tell you that it knows exactly what *you* need to make *you* love yourself—convincing you that those needs can be met outside of you. That this lost, insoluble thing is something you can gain from a better life, aesthetically. Everyone wants to be well, but few want *to do the work* to be well. This is an existential problem—we deem worth, value, on the exterior—related to the idea that happiness can be gained through material possession or the perfect body that you'll never have (but even if you had, you could never enjoy) because you were stripped of the initial confidence needed to make you feel good about yourself in the first place. If nobody has looked at you as a human, with a complex ecosystem of experiences, of traumas, of loss, of failures, of longing, of desire, of inadequacy, of hopefulness—and, on top of this, as a person with a deep need to be loved and seen, alongside an incapacitating fear that you will never get it—wouldn't that drive anybody mad? Instead, we become a target for the plight of consumerism, because we all need (and yet lack) love so deeply.

◎ ◎ ◎

Audre Lorde writes in *The Cancer Journals* about losing one breast to a mastectomy after she was diagnosed with breast cancer. She speaks of grief: "I want to write of the pain I am feeling right now, of the lukewarm tears that will not stop coming into my eyes—for what?"[6] Is there anything more relatable than a woman's worthy articulation of her pain, and then her own dismissal of said pain in one breath?

I used to think I write for the legion of sad girls, but since writing this book, I've become clearer on my mission. I write for those desperate to understand the faint hysterical patterns left like an imprint on our souls. I can't look at someone's sadness, their illness, or depression, without tracing it back to the research and findings in this book. I don't want to pathologize, but I feel that our sadness, depression, and illness are a portal to understanding ourselves, our pasts, and our families with more searing clarity. It's a door. Sure, sometimes it's one to oblivion, but my own process of allowing myself to go mad, to leak out and stain the floor, to die and die again, has led me to an incredible awakening.

Though there are still some days I resent having had to save myself. I hate how I've had to finance my own healing on a writer's salary. I've *been* mad that I had to do this work in the first place, but I also know that if I hadn't, it would have meant actual death.

In the West, there's no space to rehabilitate from a hard life. You were in a car crash and now you're paraplegic? That's your problem. You were gang-raped in college, failed all your classes, and were forced to drop out of school? That's your problem. You're neurodivergent and are confused by people because it's hard to engage with reality when there's no truth in society? That's your problem. There's no space for the most vulnerable to heal. If you have money, you might have more resources, but healing is always going to be limited because there's no after-rehabilitation process, as in a step-by-step guide to reintegrate us into society. Society isn't safe. In a city like New York, where I lived on and off for over a decade, that's a known fact. It's a hard city that relies on the continuation of hardness, which is misread as strength. But I think it's really just a denial of the needs of people. A society can't be a one-size-fits-all concept, because this always prioritizes those who have power, and I mean that in a racially, able-bodied, and class sense. When society cannot protect all of us, how could any of us think that we can feel safe or cared for?

"What if depression," Ann Cvetkovich writes, "in the Americas, at least, could be traced to histories of colonialism, genocide, slavery, legal exclusion, and everyday segregation and isolation that haunt all of our lives, rather than to be biochemical imbalances?"[7] What if, before the trauma ripened in people, we gave them space to grieve, to share, to connect, to speak the truth of their existence? What if we allowed people to be honest about how much they hold? What if we met them there, at that wounding, and didn't isolate them? According to writer May Sarton's findings, "care has its roots in the Gothic word *Kara*, meaning 'lament.' The 'basic meaning of care,' then, Sarton notes, is to grieve, to experience sorrow, to cry out with."[8]

It's interesting to me that to care can mean to lament when society offers so few places to do so. How many people ask you how you're doing and care about the answer? It's been hard to find people who can really nurture me on the daily, on the regular, who want to do that work and meet me at my grief, knowing that it doesn't usurp theirs, knowing that there's always a space for all of us *to grieve* and that multiple people can be present in those needs being met, that we can all be that for each other. Often when I've wanted to express sadness or a complaint, it has been explained back to me, or I was glass half-filled. Something stuck out—always—and rarely would anybody want to hold my pain with me. If they did, it was usually trauma bonding,* and those interactions cascaded and fizzled long before they got any ground. When "being present" doesn't feel like a worthy offering, what does that say about the society that adheres to such dissonant standards of connection?

Other people's dismissal of my feelings made me accept, again, that my needs *could* be dismissed, and that *that* was advisable. It was less messy, and it made me more pliable to others, which is what I thought I wanted. But so much of people being incapable of meeting you where you're hurt is due to their incapacity to meet themselves in those very places. Perhaps they, too, have been dismissed too many times. There's a sanitized way we meet each other, with zero compassion, and severed interest or concern beyond anything that is mercenary. It is not surprising that a society that runs on commerce would proliferate a cycle of people incapable of meeting each other on a deeper level.

And, yet, if I meet myself, if I am holding my own pain, naturally it's easier for me to hold yours as well. I'm not waiting to be saved by someone else's validation. I'm validating myself. Writing has been that journey for me, and through this process, I've made some of the most intense connections of my life.

* Trauma bonding defines a psychological phenomenon where those who are abused form an intense, unhealthy relationship to each other due to the false intimacy created in their shared or similar experiences.

Virginia Woolf, in her 1930 essay "On Being Ill," observed that it was strange "that illness has not taken its place with love and battle and jealousy among the prime themes of literature."[9] Literature, Woolf believed, concerned itself too much with the mind while pretending that the body was "negligible and non-existent." Woolf called for a new literature, one that would report on "this daily drama of the body" during illness. She believed that looking illness "squarely in the face" and writing about it would require "the courage of a lion tamer; a robust philosophy; a reason rooted in the bowels of the Earth." For these works to be penned, Woolf remarked, writers would need to invent a new language—"more primitive, more sensual, more obscene." And people would have to "reorder their hierarchies to include illness (along with jealousy, for example) among life's central experiences."[10]

◎ ◎ ◎

A few years ago, when I was living in Montreal, I remember envisioning the life that I wanted. At that age, so much of my existence was shrouded in denial, existing side by side with some uncomfortable ghosts. The life I have now has definition, despite at times being difficult to bear. But I would take this life with its continuous openings and blood-curdling revelations over the silent rupture of not knowing. I'm the generation that wants to know, I came here to hunt. Not to play a dull solipsistic game of feeling sorry for myself for a millennia only to curse all the generations to come with poison because I couldn't be fucked, *like the generations before me,* to really change. I don't blame them. Partition, famine, civil war, gang rape, the traumas that exist adjacent to genocide, monsoons, malaria, dengue, immigration to two different white settler colonial nations . . . I forgive my family for not being able to think about me. But imagine what it would mean to inherit all this?

In Montreal, I had no idea who I was, but I knew I was in immense pain. Pain, I believed, with enough insistence, I could hide. I believed I was only likable if I became so small, like a particle of dust. I was a people pleaser. I was amendable, malleable—like gold—but I didn't think I was precious; instead, I told myself it was my responsibility to placate any situation with the skill of a precise moderator. I am immensely sympathetic. I can almost always see someone's side, even if it means I have to apologize or take accountability. My skill is how well I can believe I am the bad guy and that I should apologize or submit to something I was accused of doing. As a child, accusation at my mother's hand was common. Sometimes she was right—I stole from her shamelessly. She always stole it back from me; she's stealthy, but sometimes I stole lying to her face. This gave me a complex. Was I inherently bad, and that's why I was being punished into eternal damnation by having a mother like this? Or was I damaged, broken, believing I deserved some sweet redemption that, in lieu of a mother, I stifled through food and a mild shopping addiction? Most days I was the latter, but I punished myself believing I was the former. The more I made myself invisible, playing into my own abstraction, believing in my mother's mythology of how we were all bad, I continued to lose my authentic sense self.

After my suicide attempt in the bathtub in 2015, the thing I haven't said before is that I thought of my mother—that's what pulled me out of it. The sadness that I might not get a chance to forgive her. I wanted to absolve her sins, but at that point, I hadn't even named them. There was too much to account for and not enough of her sanity to hear me, to face me, to tell me why. I will never get over the grief of not being able to have a real conversation, one where I'm not clawing at teeth for an answer from the person who birthed me. The saddest thing is that I want to know her, but she won't let anyone in. She would much rather live in a fantasy where she is the only victim of life's cruelties. I used to want to believe that it was true. That what had happened to her was so unlivable that she turned out the way she did. I always humanized Disney villains because I had my own.

On a trip to New York many years ago while I was still living in Montreal, I bought a little red book from a shop in the West Village with the word *Ideal* on the cover stamped in gold. I still have this book, and years later, I still write my dreams and aspirations, with the new moon, in there. My first entry, etching in my needs, I conceptualized my ideal life: regular massages, acupuncture, and an Ayurvedic doctor. A decade later, I have finally achieved the inherent rhythm I sought back then, in my early twenties, when I was just starting to understand the edges of my body. It's humbling to know that the self-hate wouldn't just dissipate, that I would have to begin to confront it. I knew it had something to do with needing to love myself. Throughout the years, I'd notice that I felt the most embodied after random massage work. The regularity of bodywork significantly helped me work in the journey of liking myself—the somatic healing surfaced a deep reverence I had never had for my body. It's as if the touch of someone else's hand was a way to soothe me into existence, a way to counteract all the violence that hit my body. I had often felt too shell-shocked by the world, because my initial entry point into life was so much devastation that I was forever in the folds of the fear that beckoned me as a child. To get out of that trauma state, or to not always adapt and fall into that trauma state, requires continued resolve. I started to also see how many of us believe things like acupuncture or massage were extra, and therefore unnecessary, as we've silently taught to be callous with our own bodies, and thus somatic needs. I began to see how much shame there was in claiming care, or even stating a need for it. We have all accepted the failure of the status quo . . . but what about the bodies, like mine, where this lack of care could be lethal? For me, the work had to be done, otherwise I'd die. I think I would've killed myself from the pain and the misery of physical discomfort. I had no choice but to begin to heal somatically, which required a regularity of touch, of care. During the pandemic, this was immensely difficult, but the clarity of the time gave me conviction to claim myself more ardently. After a conversation with my friend Jamie, where we acknowledged our own chronic body pain, we began to dream of the possibilities of affordable care. What if everyone could afford a weekly massage? Weekly acupuncture? Why weren't these things more accessible to those who needed the most, *across class lines*? Why is caring for yourself your own responsibility? And, in the revolution, wasn't there immense possibility to re-envision what real care could look like, *for all*?

I didn't understand that an expert therapist could know how to unravel you slowly so that you could begin to see yourself less fragmented. Early on, when I kept telling E that I wanted to be better, she asked if what I really meant was that I wanted to be whole. I knew there was a lot that was in the dark; I knew that from the signs of death and decay on my body. There have been many times in my life that I have felt hanging between portals, life or death, who knows—the underworld, or here—and I was spending more time in a deep psychic space, those treacherous planes. At a certain point, I realized I didn't want to be sad anymore. I had to understand how to harness the nectar of life, the one I had tasted, that I knew existed. To save myself.

You know how enforcing a good habit or breaking a bad one makes you feel alive? Well, self-care feels like that for me. A chance to choose myself again and again. For so many of us, accessing ourselves means breaking down the tunnels of cement we've placed between us and our own internal knowing. Some of us as children are punished when we ask for care, and many of us are admonished by our parents in times of need. Our desires are met with annoyance as opposed to openness and gentleness. Every child deserves to understand that their words and requirements have meaning and value. When you are made to feel small, or punished, especially when you ask for something, then you begin to absorb everyone else's needs, just like you've taught your skin to harden so you don't feel the hits anymore. Just like you've dulled your nervous system so that shrillness doesn't penetrate you anymore. Numbness is a survival tactic.

◎ ◎ ◎

Learning the value of shared reciprocity in relationship, that there is a shared burden and that each party should always be considering the needs of the other, as well as themselves, has been a transformative recognition. I was socialized—because of my gender, but also as a South Asian, Bangladeshi woman—to do everything, to be silent, for everyone. Asian labor across the colonized world is the labor of caring for others, and is thus seen as feminized labor and therefore undervalued. Much of my parents' decision to leave was because of the violence in Bangladesh that was making all of our livelihoods dangerous. Caring for myself, when this is my legacy, is, as Audre Lorde tells us, an act of warfare.

I wish I had something *more* profound to tell you about my actual practice of self-care. I can tell you I prefer deep tissue massage. That reflexology is king. And that the kidney 1 acupuncture point (with the needle) is like a painful but relieving sex position. I could tell you all salads are not made equal (and they're not good for you all the time depending on your Ayurvedic, TCM composition, or digestive system). I could tell you it's boring if you don't know what ashwagandha or reishi or turmeric actually do—and even more boring if you don't support decolonial spices* and only buy from white-owned companies who don't give their employees healthcare, lol. I could tell you that if you don't know what self you're caring for, it's harder than if you did. So, in that sense, what I can say about self-care is that the best of it happened when I decided I was going to pay attention and validate every little thing someone else had told me to distrust and dismiss about myself. Last year, during a moment of prayer at my altar, I noticed that my foot was cramping about four times before I moved it. When you're always on fire, what's a little more fire added to the situation? "Sick Woman Theory," writes joanna hedva, "redefines existence in a body as something that is primarily and always vulnerable, following from Judith Butler's work on precarity and resistance."[11] As humans, our mere existence on this planet is precarious. So, what else do we have but each other, through this?

I hid my chronic pain starting in my mid-teens, believing that the pain that I endured was my path, a path I had seen my mother and the other women in my family take. One of martyred servitude. But when I started to look at myself more intently, with no narrative attached to my own being other than the notion that I wanted to

* The original intent of colonial conquest of the Indian subcontinent was a desire for domination of the spice trade. Founder of Diaspora Co., Sana Javeri Khadri, an Indian woman, created a way to counteract extractive ways of getting Indian spices by working equitably with farmers. "About," *Diaspora Co.*, https://www.diasporaco.com/pages/about.

know myself better, I released myself from the servitude. Now I was something worth knowing, even if it was just to myself. I wanted to understand the shattering, I wanted to see the glass on the floor, I wanted to see where the rock landed and determined the real gravity of its impact by finally facing who threw the rock and why. At a certain point I began to see that this recognition of the multiple stories, of the multiple paths, of the endless realities, were what made up life. The purpose of being alive wasn't to be right, it was to accept and understand yourself and others and compost whatever wreckage you could do to heal the cracks, for yourself as well as the generations to come.

"Torn apart into basic elements and then reconstructed," writes Gloria Anzaldúa in *Light in the Dark*, "the shaman acquires the power of healing and returns to help the community. To be healed we must be dismembered, pulled apart. The healing occurs in disintegration, in the demotion of the ego as the self's only authority."[12] I realized I had a responsibility to myself, and the clear indication was that my body was sick, that I was unwell, and that I needed extra attention from myself. For the first time, with a sense of relief, I realized that maybe I was beginning to figure out how to take care of myself, and maybe all I needed to do was listen to myself—to my body—and trust myself. For the first time, I made that connection without shame. A lot of us might have the switch—the red light that goes on with the same kind of interior recognition—but to trust oneself is also a journey of self-care that consists of rewriting, redetermining, and rerouting how you exist in this world. That commitment is what I believe self-care really is. It's what Estés says is "becoming alert by oneself, for oneself. Letting die what must die."[13] It's remembering you're all you have, and that there's grace and real love in accepting that.

CHAPTER 10

ON SELF-CARE AND SELF-HARM

My default setting is primarily self-harm. When I look back at my twenties, I see someone desperately trying to understand who they are but failing because of a lack of tools and resources. My tendency to self-annihilate, I assumed, was a strange sado-masochistic quirk that I downplayed as a charming attribute. You know how people cloak things in archetypes, like an "artist's life." As I began to develop race consciousness, politicizing myself, at first I assumed it was the plight of being a person of color in a world of white supremacy. Upon further examination, I realized that was just one part of the jigsaw puzzle.

But when you're a woman, inquiry into the self is seen as a solipsistic endeavor. Throughout my life, that much has been evident—the way people would constantly couch my depth in language that disregarded me. We puncture and poison each other through the many ways we enforce gender. "Early in the formulation of classical psychology women's curiosity was given quite a negative connotation, whereas men with the same attribute were called investigative . . . ," Estés writes. "In reality, the trivialization of women's curiosity so that it seems like nothing more than irksome snooping denies women's insight, hunches, intuitions. It denies all her senses. It attempts to attack her most fundamental powers: differentiation and determination."[1]

In Judith Lewis Herman's *Trauma and Recovery,* she shares that Sigmund Freud, after declaring, in an 1896 report on eighteen case studies entitled "The Aetiology of Hysteria," "I therefore put forward the thesis that at the bottom of every case of hysteria there are one or more occurrences of *premature sexual experience*, occurrences which belong to the earliest years of childhood,"[2] privately repudiated his claims within a year of publishing the report, because "hysteria was so common among young women that if his patients' stories were true, and if his theory were correct, he would be forced to conclude that what he called 'perverted acts against children' were endemic, not only among the proletariat of Paris, where he had first studied hysteria, but also among the respectable bourgeois families of Vienna, where he had established his practice. This idea was simply unacceptable. It was beyond credibility. Faced with this dilemma, Freud stopped listening to his female patients."[3] Out of this research, Freud developed the theory of psychoanalysis. Yes, the most dominant psychological theory of the next century was founded on the denial of women's reality and testimony. "Freud had concluded that his hysterical patients' accounts of childhood sexual abuse were untrue: 'I was at last obliged to recognize that these scenes of seduction had never taken place, and that they were only fantasies which my patients had made up.'"[4]

I'm glad that the evidence is so blatant. We have constantly been denied our truths, while we are cajoled to believe that we are wrong, that we are hysterical, that we are the madwomen. All along we have been screaming to be taken seriously, while our bodies, minds, and hearts are compromised first by the tenacity of sexual trauma (usually rooted in our home life, or in our close physical proximity) and then by the ruthless dismissal of what we know to be true. "Though the identity of 'woman' has erased and excluded many (especially women of color and trans, nonbinary, and genderfluid people)," writes joanna hedva, a nonbinary person of color, "I choose to use it because it still represents the un-cared for, the secondary, the oppressed, the non-, the un-, the less-than."[5] Our stories are radical because here we are, in the twenty-first century, still having to determine and explain our humanity—let alone our trauma—to you. It's absurd that anyone could think that someone would hallucinate—or desire—this level of violence. That our fantasies are a representation of our inherent brokenness, not the consequence of the actions taken against us by the people who wish to overpower us.

For as long as I can remember, I have wanted to die. The earliest that was palpable was at age ten. At the time, we were living in Brisbane, and I was already completely disassociated from my body. I didn't know how to be *in my body*, and I was always trying to escape it, to exist outside of it. I spent a lot of my time in the physical darkness. I found spaces in the house where I could disappear, behind curtains, in the garden, in the folds of cracks. If I was imperceptible, I wasn't a burden; if I wasn't seen, then maybe I wouldn't have to engage with anybody and I could be left alone. When I was lucid enough in my teens, I would wonder why I hated myself so much, but I was too obsessed with my pain to really analyze it. All I knew is that it depleted me. In public, I was a happy-go-lucky high-achieving school role model—in private, I wanted to kill myself. Unsurprisingly, *Dead Poets Society* was my favorite film in my late adolescence. I related to Neil Perry—on some level I believed I was like him, wanting so badly to stay alive while the tyranny of my home life, and the expectations that had been placed on me, despite how I had been transgressed as a child, had left me in a state of suspended violence, constant alarm, and suffocation. The people whom I had tried to signal to help me, including my own father, hadn't. I now understand that back then what I felt was helplessness. "When trust is lost, traumatized people feel that they belong more to the dead than to the living,"[6] Herman writes.

Few people understand what the absence of maternal love feels like unless that's their life, too. How it is to feel forever deficient. How it is to feel like your very livelihood has no meaning. How it is to feel like you are a burden in every relationship you ever have because no one has ever told you that it is safe to be who you are. It is to feel a calcium depletion where you are left with brittle bones in a world that will try to endlessly collapse you. When your mother has rarely shown kindness, how do you learn to source that from within yourself? Where do you find it, and how? For me, the journey has been over two decades of searching. Searching in friends, desiring protection but instead facing unreliability, instability, and people's attacks and insecurities against my own.

I tried to find God, to call out, but God hadn't been helping, and I felt denied love again. So, art became my proxy love. I was consumed by poets. By Rumi, my Muslim brother, my kin. I would watch films and find myself in the folds of the screen, in the pages of Tolstoy's *Anna Karenina* and Arundhati Roy's *The God of Small Things*. My life was bleak, but it was also majestic in its misery. I understood myself through art because my pain was large like what I saw in the stories, in the paintings, in the opera of film. My heart has always been too big, and I have always been too eager to give it away.

I have struggled with my sexuality—both because of my latent queerness, which obviously started with and was cemented by my mother, but also my desire for sex, for the control of it, for its grip, for how quickly I can lose myself in it. In the words of Kate Zambreno, "I wanted to be fucked and stay fucked." My desire was potent, it felt like a penetrating hunger never satiated. I would consider myself an incredibly sexual person, and though I have spent many years consumed by sex in an almost hedonistic, punishing way, I have also been celibate for months and years at a time. The last stint began shortly after I realized I had been sexually abused as a child. You can only deny somatic injury for so long. Herman tells us, "Some survivors may conceptualize the damage of their prolonged captivity primarily in somatic terms. Or they may become so accustomed to their condition that they no longer recognize the connection between their bodily distress symptoms and the climate of terror in which these symptoms were formed."[7] You could say my body shutting down to sex is a symptom that I began to notice that was in such juxtaposition to how I've felt about sex for most of my life. I was carnal, I was raw—that's how I saw myself. I felt powerful, I felt desirable during the act of sex. Not something I otherwise felt. But there were years and months at a time that I shut down, my body completely withdrawn from sex. This contrast was confusing. I had so many lost parts of myself, and they all felt disparate to each other. I wanted to know what the connector was. I wanted to understand what I wasn't seeing.

There was no sense of safety from my mother's leering eyes; well into my adulthood, if I was wearing the wrong thing—I never dared, so it was often by accident—I was admonished to the point of forced compliance. I would be called a slut, a whore, a seductress hell-bent on spreading evil everywhere I went. This was so often the archetype I was forced into, and it would make me feel so ashamed, a crawling sensation of forever being vilified for the wrong thing.

With abuse two plus two equals five, so if a caretaker tells you enough times that you are a bad person, that your body, your femininity, your very existence is bad—even if it's for some arbitrary reason—you begin to believe it. "The abused child's sense of inner badness may be directly confirmed by parental scapegoating. Survivors frequently describe being blamed, not only for their parent's violence or sexual misconduct, but also for numerous family misfortunes,"[8] Herman explains. I learned my body was unsafe, even in my own home, and that the violence it met was the punishment for my countless sins.

There were a lot of things that worried me at this time, among them the cascading shame I carried with me, and how that made me want to hide myself and my body. As I got older, I realized it wasn't by accident that I kept getting into weird sexual situations where I was misreading the intensity of the interaction and convincing myself that what I was feeling was love. Then there were the countless individuals I had sex with to minimize the pain of rejection, but then would enter mercurial dimensions with them as well, never safeguarding myself, my body, my heart—not even understanding that was a choice. By the time I was back in Australia, visiting my mother on a strange half pilgrimage of healing, I was working on this. We didn't have (and don't have, never have had) a relationship where we talked about real things. I can't remember any full conversation I've had with her. She wasn't/isn't someone who checked in with you or asked you how you were, so my care was always one-dimensional. My father once pointed out that she would never even ask you about your day. For the most part, I didn't mind. I was accepting of her shortcomings and was compassionate toward her own failings and traumas that blockaded her ability to be fully present as a mother. I was eager to change gears, I wanted to forgive her.

"If the abused child is able to salvage a more positive identity, it often involves the extremes of self-sacrifice,"[9] writes Herman. "The normal regulation of bodily states is disrupted by chronic hyperarousal. Bodily self-regulation is further complicated in the abusive environment because the child's body is at the disposal of the abuser," she adds. Is this why was I always expecting to be abused, even if I was hoping for *more*? The way my mother would abandon me, us, in moments of her disgust or disappointment was deeply devastating, something I'm still metabolizing.

"The emotional state, usually evoked in response to perceived threats of abandonment, cannot be terminated by ordinary means of self-soothing. Abused children discover at some point that the feeling can be most effectively terminated by a major jolt to the body. The most dramatic method of achieving this result is through the deliberate infliction of injury. . . . Repetitive self injury and other paroxysmal forms of attack on the body seem to develop most commonly in those victims whose abuse began early in childhood."[10]

When I was quite young, I learned that physical harm, when administered by myself, had a numbing effect. I perpetually felt disgusting and lived in constant fear that people were thinking exactly that as they looked at me, that they could see how ugly and useless I was. The first time I took a razor to my leg, taking a chunk out of my calf, I felt like I was purifying my skin of this wretched hair on my unfeminine body. When the blade came down, taking off the first layer of my dark skin to show the white underlayer, now blistered with the sliced open pores, blood rushing like a deluge, there was a moment of relief. Like a sigh at the base of my throat, a release that felt karmic, divine, elucidating death awareness, a proximity I found intoxicating. I longed for death in a poetic way, as an act of penetrating vindication.

I continued to cut myself more regularly into my teens. It was literally a way to minimize myself. Take a stab at myself, the flesh wound a portal to another world. I became obsessed with the end of my mother's clothing shears, the sharpest object I knew in the house, outside of knives—but knives felt too close to the bone. After my mother's knife attack, I preferred something more subtle. Like the point of the safety pin, the edge like a crisp, sharp salutation against my skin. I liked bandaging my skin, cuts I made by accident, clumsiness—but also precision—nursing my wounds with the terry cloth bandages scavenged from my mother's first aid kit. I liked how the rub of the cloth against my wound reminded me of my secret wherever I went. My strange, fucked-up secret.

I can still feel the sensation of wanting total obliteration. It's something I have craved through sex, but also when bad news comes. Sometimes, in order to shield the landing, I would buffer myself by exerting more pain onto my body. Either by cutting or sometimes through punching myself, screaming into my pillow to cut the sound of my life crumbling beneath me. As a child, I grew accustomed to cleaning my skin under severely hot water. I taught myself to do that after being told that I was going to go to hell so often that my child self began to prepare for hellfire by preemptively burning my skin. In adulthood, I've tried to interrogate my obsession with heat. Heat was comfort—a hot water bottle in lieu of a mother, as well as a purifier of my sins. Once when I was ten, my mother, after missing my bowl, poured *hot, hot* soup onto my lap, searing my little legs with burns. When I screamed, wanting the natural apology or concern, she looked at me—bored, as if I very well should shut my mouth and enjoy my food. I went to the bathroom and cleaned myself, crying on the cool tile beneath me. Eventually she brought over potato skins to put on the burns. She chucked them at me and never apologized. To this day I'm not sure if she did it on purpose.

"Survivors who self-mutilate consistently describe a profound dissociative state preceding the act," Herman explains. "Depersonalization, derealization, and anesthesia are accompanied by a feeling of unbearable agitation and a compulsion to attack the body. The mutilation continues until it produces a powerful feeling of calm and relief; physical pain is much preferable to the emotional pain that it replaces. As one survivor explains: 'I do it to prove I exist.' . . . Self injury is intended not to kill but rather to relieve unbearable emotional pain, and many survivors regard it, paradoxically, as a form of self preservation."[11]

As a form of self preservation.
As a form of self preservation.
As a form of self preservation.

What was I doing if not preserving myself in writing, archiving my life for those who may share these traumas or twists and turns? Looking at why I so badly wanted to die has been a process, but now I understand it, fully. I understand how tricky and mercurial sexual abuse is, how unsatisfying it can be to identify it; how skin crawling the experience is to detail it, but the more I openly talk about it, I realize how many of us carry a deep shame for the way our bodies have been misinterpreted by people we thought loved us. My freedom feels like it was ripped from me. For most of my life, I have felt that as a child some part of me died and died again, causing me to believe that I was a dead thing as opposed to a pulsating ripe being ready to fight for myself. I guess I just needed to see that there was a self worth fighting for . . . and that came by loving myself.

Self-care can be an arduous journey. How do you love a self you don't? How do you care for a self you have been conditioned to fight? If we can't even face ourselves in the bathroom mirror, a mani-pedi is not going to save us. Some of us have to fight *that* self from taking over. That's what the journey of survival requires, a want to not end yourself every day. You have to find some lightness through it all. To confront the darkness, you need a hell of a lot of light, and that, too, I believe, comes from real earnest love, or with the smallest baby step, to find what works for you. But you deserve to see yourself—to know yourself is such a gift.

These days, that's what I focus on. I have to remind myself that I have always found ways to begin the journey of loving myself. The first time I got a tattoo was shortly after my abortion. A few months later, on an impromptu healing weekend, I went to Melbourne for the first time, only to get a tattoo of the slivered new moon and three stars. It was my homage to my Muslim roots, to my pride that I felt for my lineage. My next one was *La illah il Allah* in a shitty cursive I got from a dingy tattoo parlor in Sydney called the Invisible Man. Fourteen years later, I have over fifty tattoos all over my body. I'm glad that I wanted to find ways to love my body at that age and figured out how that would look for myself. My first tattoo was an absolution, it was the first uttering of me to myself, acknowledging that this body that I hated so much, that had always felt like such an ugly hunk of flesh, I was now molding her, I was molding with her. I was blossoming with her, allowing her to breathe, allowing her mercy. Looking more into sigils, I learned there was a lineage in Thailand where ancient spiritual forms of protection were etched into the body. These were prayers, as mine were prayers, inked into skin for all kinds of remedies and spells. It also cut across class lines, where all sectors of society got tattoos. In the West, they were relegated to thieves and thugs, but in Thai culture they were highly spiritually potent acts. Usually through pure geometry, prayers were as specific as the Hindu god Hanuman going into battle with a spear, etched into skin as protection; the sigils were generally designed to favor good luck and power in any given situation. What I liked most about these Thai spiritual coordinations is that their incantation relied on your moral code. Keeping the power of the tattoo went hand in hand with being a good person; otherwise, the energy would leak, rendering a capsized prayer.

"Tattoos play a significant part in the role of the body as archive," writes Tamara Santibañez in *Could This Be Magic? Tattooing as Liberation Work*,[12] articulating what I have felt about my own body for a long time. That it would speak things my voice could not. A tattoo is a boundary. It started as a shield, a marker of my pain tolerance. Then it became a way to mark my flesh as my own. It is mine, this vessel, and I know where it begins and ends like how I know the compass of the etchings on my skin. "Tattooing is basically anti-repressive. I think people's main subconscious motivation is to clarify something about themselves to themselves, and only incidentally . . . to show other people."[13]

A boundary is also a boundary, and in my evolution as an artist and a person, I am beginning to understand how easy it is to be exploited when you don't know your own self, and thus your own parameters. I'm learning every day that a boundary is knowing what is yours and what is another's. It's understanding your value is not determined by how you manage others and their needs, it's about knowing your needs are equally important, vital, to your soul—so you deserve to prioritize them, you deserve to listen to them. "Without firm boundaries," writes Bethany Webster in *Discovering the Inner Mother*, "we can easily become enmeshed with others, causing us to emotionally caretake, be overly responsible, or neglect our own needs. When boundaries are too rigid, we isolate ourselves and push others away. Healthy boundaries are 'selectively permeable.' They are not too rigid or too loose."[14] They are flexible, they are fluid to embrace the uncertainty of life.

Transmutation is one of my favorite cards in my Medicine Cards deck. It's symbolized by the snake, a snake I now have tattooed on my skin, as a reminder that I can sublimate anything into gold. I have always had that power. It's funny that only now I'm seeing it, seeing all these gifts given to me in the folds of my skin, in the voids where the trauma was once stored. These days, I rely less on self-obliteration. The instinct has dissipated. I find myself castigating myself less, too, and the tendency to want to cut myself to haze my peripheries is completely nonexistent. I can't say this is an arrived state, but it feels different. Healing is changing composition, it's moving through the tunnel of your own demise knowing that eventually the muck clears. Sure, maybe there's a whole new shit show awaiting you to face after, but what else is life but this? This beautiful, cyclical, spiral dance. Even our healing is art like a dervish's trance. Speaking of, I love that nineteen-year-old self who hated her body so much that she knew she had to find a way to love it. Tattoos gave me my body back. But on my own terms. It doesn't matter how many family members will tell me I'm going to hell, all I know is God led me here. To this body's acceptance. That is an act of God. To love what was previously unspeakable. To find beauty where only pain once permeated is a phenomenal act of faith, of grace.

CHAPTER 11

ON EROTICISM

The tambourine begs, Touch my skin so I can be myself.
Let me feel you enter each limb bone by bone, that what
died last night can be whole.

—*RUMI*

At the time of writing this, I will have been celibate for over two years. From my late teens into my mid-twenties, I identified as a nymphomaniac. I guess that's what happens to so many of us who are sexually repressed. We can get obsessive and unruly, consumed by craven ways once we taste it. A lot of this, I believe, leads to aggression within realms of sex, as well as murky understandings of consent. Sometimes, in order to allow sin, you believe in your badness because the possibility that you could be complicated was never something you were allowed to be. I wonder how much of my oversexualization had to do with an identification with badness. How else could I explain myself to anybody? I was ashamed of my actions. That festered and made me sick. Cyclically, I kept getting into bad sexual situations, I kept leading myself into the dark because I couldn't see a way out.

Obviously, a lot of this comes down to being a child sexual abuse survivor. When your body is exposed to nonconsensual desire at a young age, it's confusing to then deny yourself any arousal that feels true, honest, even if it's horrifically tied to memories of abuse. Memory is layered, and it evokes so much. It makes sense to me now, in retrospect, that I was a body constantly aching for some mythical touch, believing it could redeem me from my lust by satiating my hunger. As I sit with how my body was compromised as a child, I also think of the rampant ways that children are discarded and forced into child prostitution in Bangladesh, the lands of my ancestral lineage, and realize how the macro always connects to the micro—how the denigration of a child's body, a woman's body, is a societal sickness and a conduct that is accepted, then allowed and then fostered. Even though Muslim societies like Bangladesh are notoriously conservative, sexual exploitation within a stratosphere that polices sex to a sociopathic degree is still rampant. What does that say about Muslim societies? As Faty Badi tells Leila Slimani in *Sex and Lies* about Morocco, "Our society is still very prudish and conservative and . . . totally obsessed with sex."[1]

It's no surprise that in private, a variety of my Muslim friends—of all genders and of all races—have shared their abuse at the hands of Muslim schoolteachers, Qur'an teachers, men who are supposedly close to God but end up molesting young children. Funny how that keeps happening in circles of devout men. If it's not that, we're molested by our Muslim family members—parents, cousins, siblings, grandparents. I was always taught to fear what the outside world could do to me, which is what kept me locked in doors, where I faced the extreme abuse I was apparently being protected from.

I've seen a lot of Muslim women struggle to understand their sexual appetites because many of us were raised in spaces where eroticism is diffused, denied, or forced to be hidden. Yet it is still palpable in the way Muslims watch and surveil other Muslims. "Idealised and mythologized, virginity has clearly become a coercive instrument intended to keep women at home and to justify their surveillance at all times," Leila Slimani writes. "So we find ourselves in cars, in forests, on the edges of beaches, on building sites or in empty lots."[2] I lost my virginity to my Hindu Indian boyfriend on the outskirts of the Sydney Botanic Garden as a teenager. Just because I was told not to have sex didn't mean I wasn't going to. And after what had happened to me, I didn't feel as if I owed anyone an explanation.

By the time we Muslim women have any agency over our bodies, we are either hiding from dark pasts or denying our true desires. Needless to say, we are also not a monolith, and I don't think anything can be put into a binary, but many if not most of the Muslim women I know intimately have sexual health problems. For me, it was the constant bacterial vaginosis (BV) I kept having with one partner that broke the proverbial camel's back. Though I assumed our pH levels were unbalanced, I also later realized that my partner never went down on me and always wanted to enter me immediately, as if I didn't need arousal. At the time I wondered if part of the problem was that my body was responding to my needs not being met, but I also didn't know the language to ask. It's moments like these where I know my life force was lacking—I was frozen in subservience, my own needs far and distant.

Beauty is a privilege, of course it is—and of course, like all things—there's a darker, more sinister side to those whose beauty is spoken and defined for them. As a child sexual abuse survivor, I had only ever seen myself as disgusting, so accepting my beauty came with trial and error. It felt impossible to be both—beautiful and disgusting—one had to be truer, so I opted for the latter. I realized, by my twenties, that I had a certain kind of pull, that people were drawn to me. I wasn't sure what it was about me, but it was like honey. I was ripe, and people wanted a taste. This was lethal, as I began to feel like maybe through sex (and my redefinition of it) I could claim what I felt had been missing. I wanted to believe sex could feed me in a way virginity hadn't, and I hoped that it would make me accept myself more. In reality, the sex I had was often punishing, and mostly not in a good, redemptive way. I realized I wanted to dissolve into another human, become their flesh, saliva pooling from body to skin to mouth to collarbone, spit like a trail of hearts, I wanted to be consumed. While also being loved, adored, protected.

Acknowledging my sexuality and its dimensions was an important distinction I needed to make for my own liberation. It wasn't until I started to understand my sexual abuse, however—to exist within that prism, that knowing, that utterance—that I began to understand why I kept getting shattered. It wasn't always bad—there were times when I fucked that I was freer than I had ever been because I wasn't running anymore, I was in the pain, I was in the pleasure, I was alive with the yawn and heat of what it meant to be next to another body and have it enter you. I can tell you that I craved touch like I prayed for mercy. With a deep, agonizing scream. When I finally had sex, it was rapturous, never for the other person's efforts sweating alongside me, but for my ability to seek such heights. To be in the folds of another person felt like being home. It was an arrival. But, at a certain point, I couldn't just keep arriving and never transforming. I was delaying myself, distracting myself through the suffocating satisfaction of being plundered. I began to feel like my body itself had traveled far from me.

Being a femme person in this world comes with so much unsafety, especially when your body and you are blamed for other people's misdeeds, but instead of having solidarity against the limitations of the patriarchy, you are punished for a beauty assigned to you. When the way you were born to look becomes a cruel catalyst for an unsolvable sin, who can you turn to? Who protects children like me? Does God forgive me for the things that have been done to my body because it didn't matter if I said no?

For most of my life, I was so deep in the darkness, in the trespass, that I couldn't fathom that I could have a relationship with sex, with these human parts of me, these tangible parts that longed to be touched. Yet when I could forgive myself for desiring desire, for wanting to get lost in it, something miraculous happened—my body began to heal. It needed permission, but also an acknowledgment of the truth. My body's testimony is what brought me to the light, what brought me back to myself. It was believing that I was a complex being, and that a God who had made me so, who had given me my particular struggles as well, would forgive me. "The paradox of the situation: having so long considered women to be dangerous provocateurs, people whose sexual appetite must constantly be kept in check, we undermine the very concept of this purity we're seeking to preserve. I felt guilty before I had even sinned,"[3] writes Slimani. And yet, as director Nabil Ayouch tells the Moroccan-French writer, "We're forgetting that it's we Arabs, we Muslims, who shocked the West with our erotic texts in the fifteen century. We invented the realm of the erotic. We're suffering from collective amnesia."

And this is, as always, where the plot thickens.

◎ ◎ ◎

Muslim feminist and scholar Asma Lamrabet writes, "From the ninth to the thirteen centuries, when Islamic civilization reached its apogee, literature and the erotic arts flourished."[4] Quoting Tunisian author of *Sexuality in Islam,* Abdelwahab Bouhdiba: "To rediscover the meaning of sexuality is to rediscover the meaning of God. Sexuality properly performed is tantamount to freedom assumed."[5] Bouhdiba believed that restrictive, puritanical, gloomy view of sex was contrary to the very spirit of Islam.

To me, this makes sense. The sort of violent abstraction of sexuality in the Muslim world, or even within a land like South Asia—that produced the *Kama Sutra* and tantra—especially as I've come to find that both the Vedic texts and the Qur'an speak to pleasure, began to prove that something had been lost in translation. I instinctively understood Islam's parameters, even if I didn't always relate to them, because I understood being a person of faith also required doubt. That within the paradox, God could be found. There was a mercurialness to Islam that had been completely lost, and as I discovered this I began to realize a lot of other Muslims had a surface relationship to Islam, and didn't know much about it, outside of its stricture, which is a lack of connection adopted through the fundamentalism of Islam. Pankaj Mishra writes, in *From the*

Ruins of Empire, "Until late into the nineteenth century, people of societies with belief-systems like Islam or Confucianism at their core—much of the known world—could assume that the human order was still fused inseparably with the larger divine or cosmic order defined by their ancestors and gods."[6] Isn't it strange how an entire people's fate can change in a century?

In a colonial sense, the truth is that whiteness and the aggressor depended on shaming people whom they colonized of their culture. As Muslim historian Ghulam Husain Khan Tabatabai wrote in 1781, "No love, and no coalition can take root between the conquerors and the conquered."[7] Yet facing this shame, speaking to it, giving words to how our bodies and our cultures were othered, forcing us to seek reprieve in whiteness and the standards that those who took from us believed made us inferior—like our awareness of our bodies, our legacies of music, dance, of rapture and evolution. "True sexual liberation can actually threaten capitalism and a lot of colonial structures and therefore it's not embedded in our education system, that's why this work is so important to treat the next generations," Bundjalung woman and activist Ella Noah Bancroft says on the *For the Wild* podcast. "So much of our pleasure, or what we think of pleasure, is derived at the expense of earth."[8] This is where self-care is lost. We think caring for ourselves is actually about getting other people to care for us, or that the Earth should pay for our gargantuan needs as a society. Our inescapable greed can't be quenched by more material possessions, self-care can't be through land-grabbing or denying the realities of capitalism. Healing ourselves must happen in tandem with healing the Earth, we must understand things holistically. "The psychology of authority plays a very important part in the awakening of man's latent power," Swami Satyananda Saraswati writes in *Kundalini Tantra*. "It is not well understood by modern man who has unfortunately accepted that man lives for the pleasure principle, as propounded by Freud and his disciples. The psychology of austerity is very sound and certainly not abnormal. When the senses are satisfied by the objective pleasures, by the comforts and luxuries, the brain and nervous systems become weak and consciousness and energy undergo a process of regression. It is in this situation that the method of austerity is one of the most powerful and sometimes explosive methods of awakening."[9] A lot of the Vedic and Buddhist texts are of the understanding that spiritual growth comes with true spiritual labor. The same could be said about Islam and most other Eastern faiths.

As a young child, I watched Islamophobia rise through the media, in the streets, and across the globe. The West refused to see its contribution to Islamophobia and how it was a clear pathway to the advent of Muslim terrorism. Ernest Renan, a French Orientalist and Semitic scholar, in 1883 "denounced Islam as the begetter of despotism and terrorism. He invoked the racial hierarchy in which rationality, empiricism, industriousness, self-discipline and adaptability defined Western man, and their near-total absence defined the people he dominated," writes Mishra in *From the Ruins of Empire*. "He also stridently identified progress as the unique achievement and prerogative of the white race and Christianity, arguing that Islam and modern science were incompatible."[10] Similar to Bernier, but within the realms of sex, I like to think about how people evolve, in a social sense, when there are declarations like this smattered across history books esteemed for empirical knowledge. Yet when we start to question that very source, lots of things begin to reveal themselves and disintegrate within the realms of pleasure, and what colonization has done to the concept of desire for us the colonized. "The West won the world not by the superiority of ideas or values of religion but rather by its superiority in applying organized violence in most countries," writes Shridhar Sharma.[11] What Renan proclaimed in 1883 was almost one hundred years before Islamic terrorism was even a theory.

I was raised with idea that Muslim men were dangerous. What I saw everywhere around me, including a lot of my interactions with Muslim or South Asian men, felt uncomfortable. Yet to present men from my region of the world as the primary aggressors (how many times had a friend, usually a woman, told me that their parents would allow them to marry anyone except a Muslim man) felt like an offensive admonishment. Still the question was important: Why did so many of the women in my life have domineering relationships with men from these particular parts of the world?

When we think of the colonized, rarely do we think of sex, but looking at sexual relations can be an indicator of the tactics that were used against us to socialize us. As witnessed, the declaration of difference was first pathologized by whiteness, meaning they constructed the dictum by which we govern race. Before white people did this, we were all relatively unknown to one another, our societies were more localized. So, to say one group of people is "x, y, z" without thinking about context only exaggerates and supports white supremacy.

I think of the men from my region of the world, Muslim and South Asian men, as beautiful creatures who were Romantics and poets, artisans and charmers. This was the uniqueness of my brethren, a quality palpable in the tomes of Hafez, Kabir, Rumi. I see it in my father, in the way his thin body wails against the wind. His fragility is something I find most beautiful about him, it's why I deemed him worthy of my protection early on. But it's something I've also seen that he himself contends with, how people so easily dominate him. I think about domination a lot because colonization is a form of domination. Think about sex when you think about loss, now think about a person who loses their land to an aggressor who comes and steals everything from them, then tells them it's their fault, blaming it on their weakness.

My friend Ali once shared a story at our friend (also his partner) Salman's studio on a recent bright summer's night in Greenpoint, Brooklyn. He spoke about how the last Muslim ruler in India who lost his land to the British was seen as a failure because he was a poet and thus could not protect his people. The men from my lands were known to have male lovers as well as wives and use astrologers in the royal courts. They could speak to Socratic methods and translate texts. They were lovers. They were free. Then something happened. The feminization of Asian and Muslim men has resulted in a shame that has penetrated these communities and hardened them. People think of imperialism as a past concept, when in fact Vietnam, Iraq, and Afghanistan are modern examples of the continuation of imperialism: Bleeding oil, creating war, and devastating people to keep them in a state of trauma is the definition of imperialism. We should be extremely worried about the well-being of these people who have been forced into catastrophe again and again. I see how this demonization has impacted the men in my family, the men in my culture, the men in my faith. Our men who have been shamed into feeling their bodies, their dynamic sexualities, are undeserving and disgusting. "All Western culture has achieved is to bulldoze the modes of traditional identification and to leave the individual surrounded by troubling ambiguities and sources of conflict,"[12] Moroccan writer Abdelhak Serhane writes. If you tell people they are barbaric, less humane, more savage, that—like the narrative of an abusive parent—seeps into your soul. "The myth that the dominant elites, 'recognizing their duties,' promote the advancement of the people, so that the people," writes Paulo Freire, "in a gesture of gratitude, should accept the words of the elites and be conformed to them."[13] Speaking about Egypt soon after the opening of the Suez Canal, Mishra writes, "Until the revolts of the 1870s, British officials counted on the possibility that the Egyptian peasant was so beaten down that 'no amount of misery or oppression would provoke him to resistance.'"[14]

Because the victor writes history, they will always favor themselves. "It is not those whose humanity is denied them who negate humankind," writes Freire, "but those who denied that humanity (thus negating their own as well)."[15]

◎ ◎ ◎

There's so much I want to say, like how I long to be touched, how I have always sought embrace wherever I could—a mother's hug, a sister's patient kiss, a father's concerned head pat. I also crave pressure on my body like a bug to a lamp—I love a firm embrace, a hand on the knee, an eyeball-to-eyeball gaze. Of course, these things can feel bad, too. How brilliant that our bodies know their own safety for us. To first listen, then trust ourselves enough to know what feels good, is a wondrous gift we can give ourselves. We get to determine that for ourselves—nobody can or should be able to tell us what feels good for us.

During the pandemic, in my aloneness, I worried about myself. I worried about the impact of not being able to feel someone else's tenderness as I healed my body, but then I started writing this book and that purpose, that drive, left me feeling less alone. Understanding myself, to be able to explain it here, on this page, has been a journey. But locating, reading, and documenting has helped me see how all of this is connected, and that in itself is tantric. I think of the Zen proverb "How you do anything is how you do everything" as I consider what I am, who I am, what I want for my life. I find an erotic kind of pleasure forming and strengthening within myself, in my own body.

"We know that sexual relationships outside wedlock are common in Morocco," Slimani documents in *Sex and Lies*. "The fact that all of them must be hidden only promotes abuse and attacks on personal freedoms."[16] Perhaps this is why I write about sex: I want to liberate sex for Muslims. Our cultures were not always what they are now. "The Moroccan legislation is drawn not from sharia or any other religious source but from substantive laws inherited directly from the French protectorate."[17] This puritanical strain of Islam is a foreign entity, and we have to start asking why. I see the power of eroticism, and how that energy is the energy of what binds us, of what binds the divine. This fluidity, this power, can be found in all things and all beings, and that is the power of faith. That no matter what, we deserve connection to divine source. We know that sexual health is part of social health just as sexual safety is part of social safety. "We cannot live well when we're afraid, when we feel guilty. This is what we mean by 'sexual deprivation,'"[18] Professor Abdessamad Dialmy tells Slimani about his life as a gay man in Morocco. We have to begin to forgive ourselves and find love in God, not fear. *Bismillah al rahman al rahim*, the words that are said before every prayer to Allah. It means God is most gracious, most merciful. It is one of the most repeated phrases in the Qur'an. Yet, how have we forgotten this? Why are we so lacking in mercy and grace with ourselves and each other? Why have we become so hateful? I want my people to be free, I want my people to forgive themselves.

◎ ◎ ◎

Accepting that I was sexual, that I needed a sexual life, was understanding myself and my faith at its core. Having a spiritual life means acknowledging all parts of your creation, including the parts you've been taught to disregard or hate or demonize. But if God is everything, shouldn't these things that we've been taught to disregard—but are a natural part of life—be accepted, loved, or at the very least seen as a vehicle for God? Everything can be an active prayer, so why is sex any different? Vedic philosophers saw gateways to enlightenment through tantra, and Sufi dhikrs through poetry to Allah, the divine, the source of everything. Were the exaltations of Rumi about God or his lover? Do they have to be one or the other? We have to stop thinking that we are removed from that source. If we all believed that we, by virtue of being alive, have worthiness in the eyes of God, and if we saw each human to have that same worthiness, then we would be a different people. It's so simple, how enlightenment works, and yet . . .

Why is such a natural part of human nature an issue of dominance and control? They say sex work is the oldest profession in the world, so don't sex workers deserve more respect when clearly their demand is high and their work is esteemed? Why has their work been relegated to realms of darkness and disgust when most people crave to fuck? Would bridging the parts of us that we've separated as "good" or "bad" help us understand ourselves more wholly? Would that then allow each of us our particular path to wholeness, too?

Acknowledging the sensual was seeing the world as an erotic ecosystem that needs the balance and consideration of all species. It brings me back to a point of my own cultural origin, which Roberto Calasso explains beautifully: "This is typical of a language like Sanskrit that does not love the explicit, but hints that everything is sexual."[19] That's why knowing and naming boundaries, of being invested in goodness and accountability, are vital to retaining the aspects of the old world where bodies were understood and accepted. I'm not saying that the past was perfect (obviously) but that at the very least there seemed to be a merge of the spiritual and the profane, there was communication between those two parts of existence. The existence of eros is fundamentally a cornerstone of human evolution—we procreate, but we also lust, and that is an integral part of our spiritual governance. Dividing the two has created a dissonance that's been hard to weave back together. When people are embroiled in shame, it's easy to forget the tether of divine presence.

Touching the earth has healed me in ways I couldn't have anticipated. I mean this in the day-to-day sense, of being around nature. The act of composting, too, of being with the dead flowers, the acrid way the white-rimmed green fungus can create a ripening among newly bought yet already rotting tangerines. Touching dirt with my hands has created a newfound gateway to understanding myself within the mercurial spectrum of life and death.

I imagine we want it all to be sterile so it doesn't make us confront our mortality, so we constantly can avoid death. "The essence of consciousness is being with the world,"[20] Freire tells us. Being with the world outside of commerce, outside of transaction, what are we to each other? I've always dreamed of living on a plot of land with my friends, each of us in our own little portal. Those years ago in Montreal, my friends Zoë and Sophie and I would dream of tending to the land, a desire that's been reflected in multiple relationships throughout my life. A desire to return to the Earth, to be with her, to feel the soil in my blood, the blood in her soil, toiling together, we are one. We are the children of clay, children of Earth, borne of her. Why are we so divorced from her? From the vastness of her ecology and the light that exists in the unique flora and fauna of this vast and beautiful Earth? Nature is erotic. To watch animals is to witness wisdom, to see how every other species on Earth has a cadence with each other, there's a dance—and even among alpha species there's a respect for the creature you kill. You take only what you need—animals understand there's a balance, there's an equilibrium. They're superior to us because there's no delusion of ego. We've propped up humanity as the apex of civilization when all we've managed to do is burn the whole thing down. It's so fucking embarrassing.

These days I understand erotic to mean balance, to mean an energy that is harnessed and pure and deep. But eroticism also means an awareness of death, it's understanding that all of life must be appreciated because all of us have a pulse, and that in itself is the erotic duality of life. How quickly death can come, and all we can do in the interim is to understand, locate, communicate, and then enjoy the pleasures that we seek.

I'm tired of pretending my body wasn't compromised by someone who espoused my safety while simultaneously demonizing my body and sexuality when I should've been allowed to be a child, to be innocent. I was first accused of being sexual while I was being *sexualized*, so what does that say about me and the legion of Muslim children who have been compromised by a so-called religious elder? There's no use in demonizing us when you desire us in the first place. If we actually acknowledged our need for sex and the deep importance of that in our society, I'm assuming so much tight-lipped perversion wouldn't exist. Ironically, better communication about sex shows an aptitude and respect for it. Imagine if we had reverence for the sacrament of sexual connection. Imagine if we actually put honest attention there?

What I want so much from touch is to feel safe. I want that from sex, too. When I close my eyes and dream, I see the immense possibilities of our future. I see how a sexual revolution, accepting and having compassion for our sexual proclivities, meaning we need to prioritize the voices that have been silent—women's voices, disabled, trans, fat folks' voices. If we let all of us speak, if we prioritized the voices of survivors of sexual harm and trauma and allowed them to pave the way toward a collective future of safety, I think we would get closer to true revolution.

It's not about accepting human nature; it's rewriting the stance. It's understanding the divine is everywhere. Eroticism is that embrace. It's about knowing that there is so much more to the erotic than sex. It's a way of life, of existing in a state of liminality, of understanding that the sacrosanct has its own erotic compass. If we accepted our ways, without demonizing them, if we understood our complexities without judgment and took effort in finding peace with ourselves, we would find true, unadulterated liberation.

CHAPTER 12

ON DIVINATION

I don't know why certain people get pulled to (and by) the esoteric. I might explain it astrologically—how if you have eighth or twelfth house placements, you (potentially) have a void or darkness or openness (or all three) that seeks the depths of the occult and beyond. These placements in traditional astrology perhaps also point to desiring information in a nontraditional way. As someone who lives and reads between the lines, I want to hold all interpretations possible. This has meant that for most of my life I have struggled with authority and people telling me what to do. I hate rules, but I also endeavor to understand them.

As a Muslim, I struggled with what parameters God had ordained restrictive. I knew as a young child that "fortune telling" was a big no-no, but everything felt cast in this weird moralistic perversion that I find religious doctrine often spills into. Everything has become dogmatic, and the fluidity of the divine is now assigned to a fear-mongering God. As an adult, I find it quite interesting that a faith and culture that has a concept such as *nazar* (evil eye), and has built an entire industry around protection from evil, doesn't do more to cultivate good, positive energy and divine magic as spiritual praxis toward Allah.

Even still, I found grace in the word *Muslim* and how it meant "to submit to God," which I believed meant accepting the tide. In moments of spiritual loneliness, I clung to the concept, especially during the pandemic, feeling there was something sacred in learning how to be, how to simply be and embrace it all without fear. To become a Muslim was to morph closer toward God. Yet the journey to God is such a personal experience. Something that's always moved me about Islam is that there is no mediator between you and the Lord, and that only you, or God, could determine that question. No one has the right to determine who is of more faith, and to me stories of the Prophet Mohammed always revealed the value in accepting people's diversity. As a result, I've come to understand spiritual practices as a way of cultivating self-awareness and self-compassion.

I've been studying astrology for roughly fourteen years, but my interest really piqued in my twenties when I began to revisit my astrological chart. When I was eighteen, Alia, a Turkish Australian Muslim astrologer, was the first person to read mine by a tableside ripe with dried apricots, spongy halloumi, green pitted olives, and a variety of baklava. My sister was with me, my buffer back in those days, as we drank bitter Turkish coffee (mine with a slurp of honey) and listened to Alia tell me some of the most devastating truths. I remembered, after beginning (and finishing) the process of writing *Like a Bird*, that by eighteen I was worried that what I had sworn to do with my life (become a human rights lawyer) was not possible given my temperament, and what I really wanted was to write. Even at that age, I realized I was not tameable, meaning: I didn't want to be. I refused to be what anybody expected of me, and I was beginning to realize that I had no other recourse but to completely unleash. I had started feeling this defiance move through me, a new factor of my personality, and I was fashioning myself as a person who said no, who made choices for herself—even reckless ones. I wanted to be free from my mother and from this peril of living a half-life when there was a whole-ass person bursting at the seams from deep within me, a person, I knew, who could easily die given the right opportunity. I knew there was enough velocity running through me to swing me into the ocean.

I remember Alia explaining my Cancer Rising, Cancer Moon, Chiron in Cancer in the First House—thereby expressing my extreme emotionality, as these positions were in complete opposition to most of my other planets in Capricorn in the Seventh House—including my Sun, Mercury, Saturn, Neptune, and Uranus—which made me, to say the least, ambitious but very obsessed with connection, communication, and cultivating higher relationship. I was stunned. My Venus Aquarius in the Eighth House finally revealed my aloofness but deep passion, my Scorpio Lilith, Juno and Pluto explained the sexual undercurrent that I had long awaited an explanation for. Then, the seal: Three different placements pointed toward writing as the best profession for me. It was the sign I needed. Something I've never expressed before is that astrology gave me the strength to pursue my writing career.

It's strange how things can strike such resonance, but I guess in a way I had needed something higher, something indisputable as the stars, to bear witness to my heart's desires. If I didn't have a parent to validate me, maybe my chart would. I had not known it was possible to listen to that part of me, but understanding my chart gave me courage. It made me realize that I had a dynamic sense of being and that I deserved to pursue what I knew myself to be. I could finally see that there was a future ahead. The stars made me go on; they were a road map to my becoming.

The older I got, the more I learned that there was a lineage of this spiritual quest, as my paternal uncle, my father's youngest brother, was a numerologist and used crystals and gemstones to help direct people's particular callings. It made sense that since my early teens I had been thrust into a world of auras, Reiki, and other alternative healing modalities because of my sister, who was now becoming a coach and healer in her own right. My closest friendships were also representing these quirks, with friends channeling aliens as a means of healing, others exploring the teachings of gnosticism, and some the dharma.

So, astrology started to become a template to understand not only myself, but my parents and other family members. I began to chart lovers and friends, and through the years of understanding my own placements, I have seen that there's a deep correlation between the stars and who I am instinctively. I know naysayers and skeptics will question astrology without really understanding how deeply scientific and mathematical it is, and like all sciences, astrology ensures evolution through literature, through ongoing peer reviews documented through dynamic conversations that span history, cultures, and faiths.

◎ ◎ ◎

I had always known of Muslim ties to astronomy, but in an abstract way. Under the colonial gaze of the West, it is presumed that the master is the ultimate seer and knower, that objectivity is the ultimate test of legitimacy. This is why it's so important to look at the margins of history, to ask why it is that some knowledge is prioritized and other knowledge demonized? Who is served by those decisions?

The word *astrology* comes from the early Latin word *astrologia,* which derives from the Greek ἀστρολογία—from ἄστρον *astron* ("star") and -λογία-*logia*, ("study of"—"account of the stars"). So, it makes sense that Muslims, like other civilizations, were in the business of self-discovery. When I found out about Queen Buran—a Muslim who was reportedly one of the first known female astrologers—I felt something click. She was married to Caliph Al-Ma'mun, the second Abbasid Caliphate, who was also a student of astrology, and suddenly a lost history was shown to me, a path to understanding myself. "We must remember that the early Arabs regarded astrology as a branch of legitimate Greek science rather than mysticism or magic," historian Kenneth Johnson writes of Buran. "As a member of Baghdad's intellectual elite, she was involved with the creation of a famed astrological academy, and she was roughly contemporary—and probably acquainted—with the best known astrologers of the Arabic period."[1] It took me well into my thirties to excavate the Muslim history that was tethered to my own reality, one of spirits and *jinn*, one of portals of energy that I couldn't explain. From a young age I was immersed in another world, another dimension completely. As an adult, I struggle with how much time was lost believing that I was wrong, that there were things I didn't understand when the answers to my own human existence were right in front of me. Now I didn't need to convince anyone—I had Queen Buran's legacy to resolve me.

As someone who has been embedded in astrology for well over a decade (before this recent astrology boom), I have been on the receiving end of obnoxious comments from those who scoff at it and question my intelligence as a result. I resent that so many people I love have contributed to my smallness by criticizing this practice, as if being an intellectual, a writer, meant that my proximity to what was deemed spiritual loftiness somehow made me a less reliable researcher. This book, in many ways, is a response to those critiques. I believe not only that our obsession with the Western sciences is a disservice to the lineages that many other systems and spiritual sciences are rooted in, but it's also deeply paternalistic and racist to assume that Western resources are somehow impenetrable just because they're the status quo.

As Jessica Dore writes in *Tarot for Change*, "While the framework of evidence-based practice is built on the assumption that the only legitimate evidence is that which has been gathered through the scientific method, I'd argue that when something stands the test of time, as symbols embedded in the tarot have, that's proof of a certain kind of efficacy, as well." Dore explains that though tarot is seen as a spiritual practice more than a psychological one, psychology is inherently spiritual. Speaking of Carl Jung, she explains his theory of the collective unconscious, which is really about a "shared psychological inheritance" that is experienced through dreams, myths, symbols, and fairy tales across cultures. "Today, Jung might have been seen as more of an artist or philosopher than a scientist," explains Dore. "His exploration of mystery through the fields of philosophy, anthropology, myth, religion and spirituality informed the development of his style of analytical psychology. And he genuinely believed that the symbols found within old stories, astrology, and even the tarot contained keys to understanding the psyche."[2]

Why is wanting to understand yourself by utilizing methodologies that aren't purely clinical, that rely on your own intuition, your own conceptualization, without just wanting or allowing someone else to determine *how you feel*, why is that so disconcerting to so many people? Energy healer Donna Eden writes, "Becoming civilized is, to a large extent, learning not to do what your body wants you to do." Dore adds, "In our culture we are conditioned to prize intellectual knowing and neglect intuitive or body knowing. We are taught that science alone is real and that other ways of knowing are fantastical or unreliable."[3] Referencing tarot reader Rachel Pollack, Dore explains that "the root of our dualistic thinking is a fear that we don't know ourselves. We see things as either *this* or *that* because we don't trust ourselves to flexibly navigate the contingencies and the in-betweens and the constant flux that is life and to still be okay."[4]

Knowing how to decipher who you are, by looking at it all, by developing spiritual practices that help you decode yourself and the information that rests in your body, helps you actively show up as a whole version of yourself. By making it a priority to avoid projecting onto, hurting, triggering, and traumatizing others—because you've identified all of that in yourself—you become more aware of your actions. Naturally this means that there is more room for accountability. "Stories unshared don't disappear; they return in relationships, silently taking prisoners. If the trauma remains unknown, unspoken, and unconscious, it does harm," writes trauma expert Laurie Kahn in *Baffled by Love*.

Coming into my own spiritual power, being able to fully see my power in moments of quietude, woke up all those things that were long dormant. Recently, I've found solace in the Vedic scriptures' depiction of kundalini energy. Swami Satyananda Saraswati writes, "By activating kundalini you may become anything in life. . . . And when total awakening occurs, man becomes a junior god, an embodiment of divinity."[6] I think of the preparation and dedication it takes to live a deeply spiritual life, and how cultures that were corrupted lost not only entire spiritual ontologies but practices, stories, lineages, and rituals tied to them as well. Who retrieves that information? I've always found it absurd that we live in a world where it's normal to suppress your magic; in fact, it's preferred. What does it say about a society that wants its citizens to keep their hands tied behind their backs, casually trusting that the powers that be have their best interests at hand?

Johnson writes about Queen Buran: "According to the historian Ibn-Khallikan, Buran 'used to lift the astrolabe and look at the horoscope of the caliph al-Mu'tasim.' One day she noticed that the ruler was in danger through some sort of wooden instrument. She sent her father to the palace to tell the caliph what she had foreseen. Mu'tasim may have been a fundamentalist, but astrologers were still taken seriously; when his servant arrived shortly thereafter with the caliph's comb and toothpicks, al-Mu'tasim ordered him to try them first. No sooner had he done so than his head or face swelled up and he fell dead."[7] Reading it, I mourned all the lost power of where I come from, where Muslims come from, and what we once were. In my mind, my journey is to help restore this information. It is to inch closer to something that has been denied to me, to my brethren. I find my liberation through these spiritual compasses. I'm gaining back my lost power.

Symbology was prominent in the culture of ancient Egypt. The Mayans had a supreme connection to the land and therefore the divine, and through that there was a conceptualization of something greater. Indigenous peoples in North America have spent thousands of years contextualizing connection to the sacred elements of Fire, Water, Earth, and Wind. This correlates to the signs in astrology* as well as the suits in the tarot (wands are fire; cups are

* The fire signs being Sagittarius, Aries, and Leo; Water being Cancer, Pisces, and Scorpio; Air being Aquarius, Gemini, and Libra; Earth being Capricorn, Virgo, and Taurus.

water; swords are air; pentacles are earth). This also has crossover in both Chinese medicine (with the addition of a fifth element, metal) and Ayurveda, proving that a number of cultures were elementally constructed. If we all had a better understanding of the elements that govern this Earth, we might have more compassion toward ourselves. What else are these divination and alternative practices but a way to engage with ourselves more intentionally, deeply?

Astrology has taught me so much about self-love and self-care by giving me a foundation to examine myself on my own terms. I've learned about my quirks through studying my chart, and accepting my contradictions (which the planets explain) has been paramount to my self-healing. It has become the entire template for how I engage with the world. I see both humans and this planet through an astrological viewpoint. Conjunctions, oppositions, and trines in your chart can shed light on the intricacies or nuances of your personality, about why you hold trauma, or who your parents are and how that's affected you. Spiritual people are too often relegated to being woo-woo, as if we all just sit with a crystal ball with no scope of rationality. Why is magic seen as an interpreter of lies, instead of another context that can offer value, judgment, and credible, resourced explanations of self? We have to question why we negate everything that we don't immediately understand. I think about Cartesian philosophy and the voices of white Christian men who determined separation as opposed to holisticim. We need to reinterpret what wisdom, knowledge, and expansion of thought truly is.

In ancient Greece, the Oracle of Delphi was said to be established as early as 1400 BCE, and was the most sacred shrine built around a holy spring. The Pythia, the high priestess of Apollo, was known for interpreting divine messages and providing prophecies under possession. People, including intellectuals, would come from all over and debate the messages, as inquiry was common and a part of the ritual. This reminds me of the Muslim salons held in the House of Baghdad by Buran and Al-Ma'mun, where dialogue was encouraged and deeply valued as a site of not only spiritual but also intellectual discourse. Yet the divine is seen as a hoax, now everything far from God is revered, sanctified. Now capital is God.

Every day I meditate, pray, and pull cards from my three separate tarot decks—the Rider-Waite deck, the Thoth deck, and my Medicine cards—to bring clarity over my own psychosis. This practice always connects me to God and spirit. Jung understood how powerful symbolism was, whether in dreams or in tarot. He was able to gauge how we think, how we feel, through symbols. Looking at the art of Ithell Colquhoun and Niki de Saint Phalle, I understand how deep this legacy of the esoteric is for humans, how comprehensive and attuned our societies once were, and how so much of who we are exists in the stories we tell ourselves. The art of divination are the relics of those lost histories. We were once far more connected. To each other, to source. All this was once lost, but the archive is declaring itself again. In *Space and Place* geographer Yi-Tu Fuan writes, "Man is the crucial and central term in the astral cosmos. He contains within him the distillation of the whole astral system. The union of astrology with the body arises out of the need to unite the multiplicity of substances in the universe and out of the search for parallel wholeness."

There's something big about unlocking untapped power and believing that the future is possible. And it starts with us naming who we are, by stating the things we know to be true about ourselves. But it's also about situating yourself in the cosmos, it's about understanding how you are participating in the ecosystem of this great galaxy, of this huge human saga that we are all in together. We are a part of something big, isn't that extraordinary? It's to remember the big and the small of it and hold those realities both together. Imagine if we could all live with that awareness all the time. Imagine if that's all it took to be present—to sit with the awe of being alive. No matter how painful and hard it is at times, there's beauty. In nature, on this planet, in the vast galaxies that separate us, the legion of stars and asteroids. There's immense profundity in our human experience that is an inherently spiritual experience just by being alive, and here. Right now. We're here.

There's a bigger mission that's asking for you to unlock your own magic. It's asking you to find your spiritual powers and prowess. If you hear the call, I hope you're listening and taking note. The revolution needs your spiritual evolution.

PART IV

INTRODUCTION TO JUSTICE

What goes too long unchanged destroys itself. The forest is forever because it dies and dies and so lives.

—*TALES FROM EARTHSEA: DRAGONFLY,*

URSULA LE GUIN

CHAPTER 13

WHO IS WELLNESS FOR?

The thing of it is, we must live with the living.

—*MONTAIGNE*

Who is wellness for? I've been asking myself this for years, wondering if it was possible to even answer this absurdly complex existential question.

When I was growing up in Brisbane, Australia, where the wildlife was plush and sometimes punishing, I understood that nature was in charge. There is a great beauty, wild and untamed, in the heart of Australian flora and fauna. But it was also dangerous, beckoning reproach. When I say I'm Australian, people often remark at the harshness of our landscapes. We have some of the world's deadliest animals, from bull sharks to inland taipan snakes to box jellyfish. As a child, you are taught to be on the lookout, to be aware of your surroundings. I got good at snapping the earth brown snakes in half with a big tree branch. I was ruthless, my heart in my throat as I did it, but knowing their venom was mighty for me and my family, I acted as a territorial avenger. I was seven. Redback and funnel spiders were common, as were huntsmans, and there were protocols that I learned through osmosis—always check the bottom of your chair if you're sitting outside, there might be a redbelly lurking. Echidnas were often on the road, and possums ravaged my mum's garden, especially when it was bearing fruit. It was normal for blue-tongued lizards to lounge on our backyard tiling, and many times the small ones would get into the house and just stay there for weeks on end until finally finding a way out. I miss the way tamarind trees plunged neighboring streets with their poo-colored seeds and how my mother and I would stack the sour fruit in our pockets to make paste for *chotpoti, a* delicious cold mung bean dish. In Australia, I got used to being with the trees, within the wide scope of the jacaranda stumps that would shadow me in between moments of play, the purple leaves lively and punctuating. The folds of the earth have always been my sacred place.

What I'm trying to say is I was raised on and close to the land. The land and her creatures were a major part of my life growing up, so much so now I wonder why I live in a country where the divorce and severing of humans with the land is so apparent. Maybe it's because if we were living in accordance to the land then we would have to also face the bodies that are buried here, the blood that has been spilled, and how the ghosts still haunt. That's the reality for every colonial nation on this planet, then this one, the United States, also has the added history of hands, of labor, of toil, of more blood, of injustice laced in her soil, in her cotton, in her roots.

I've been thinking a lot about the Anthropocene recently, about how the Geological Society of America entitled its 2011 annual meeting "Archean to Anthropocene: The Past Is the Key to the Future." "The Anthropocene," Hannah Morris explains to *Penn Today*, "is a term that describes the supposedly new epoch we as humankind find ourselves in due to our own large-scale acts of environmental destruction. It was first suggested as a suitable term to denote the increasingly observable and convergent forces of global environmental change in 2000, by Nobel Laureate Paul Crutzen and Eugene Stoermer."[1] To me (and others, like Vandana Shiva) the Anthropocene started with colonization, and though the act of colonizing existed prior to European colonizers (in China, in Japan, even with the spread of Islam in Europe, Africa, and Asia prior to this), the tenor was different, the barbarism that the former exhibited was a protracted assault at dehumanization through terror and extraction. By divorcing a people from their own cultural and spiritual compass, they were dislodging us from our power, which means sovereignty, but it also means resistance.

In *The Red Deal: Indigenous Action to Save Our Earth*, the Red Nation writes, "Framing this as a panhuman problem or a problem of the species—such as the term 'the Anthropocene,' the geological age of the fossil fuel economy—misses the point."[2] Especially when "the upper one-tenth of humanity is responsible for half of the carbon emissions from consumption. . . . The richest 1 percent similarly emit 175 times more CO_2 than the poorest Hondurans, Mozambicans, or Rwandans. Twenty-six billionaires hoard half the world's social wealth that the 4.6 billion people who make up 60 percent of the planet's population—numbers that appear to get more extreme as CO_2 concentrations rise."[3] I have come to understand that true wellness means witnessing even the grit and texture of the pain, of the wound, of the mystery that lies behind it. Studying the Anthropocene while doing research for this book, I've gathered that there's a disregard for the ways in which white supremacy has caused our climate catastrophe, through the choices of colonization and onward, and now it really is up to white people to lead us to a space of evolutionary shift. This means tending to the damage. This means white people have to commit to not just environmental action, but to understand the totality of why we are here. There needs to be a vulnerability to listen and a heartfelt desire to change.

Too often, spaces of catastrophe are taken over by anger from the wrong people instead of compassion for the right people. It's interesting, then, that "early environmentalists in the U.S. were anti-immigrant eugenicists whose ideas were later adopted by Nazis to implement their 'blood and soil' ideology," writes Sarah Jaquette Ray.[4] "In a recent, dramatic example, the gunman of the 2019 El Paso shooting was motivated by despair about the ecological fate of the planet: 'My whole life I have been preparing for a future that currently doesn't exist.' Intense emotions mobilize people, but not always for the good of all life on this planet."[5] In a piece for *Scientific American,* Ray asks, "Is climate anxiety a form of white fragility or even *racial* anxiety?"[6] These are important observations and a vital part of understanding the depth of our species' climate issues.

As a Bangladeshi, I'm well aware that the land of my parents' home will be one of the first to sink into water. Bangladeshis are also (no surprise) a significant percentage of the world's labor force. The people who pay for the wealthy's greed are the people who always pay for them—through their labor and eventually through their death. According to the Red Nation, Lakotas call *owasicu owe,* the fat taker, the colonizer, the capitalist economy." It is also what Ojibwe activist Winona LaDuke calls *weitiko*—the cannibal economy," and what Robin Wall Kimmerer calls *windigo* thinking. "They're everywhere you look," she writes. "The stomp in the industrial sludge of Onondaga Lake. And over a savagely clear-cut slope in the Oregon Coast Range where the earth is slumping into the river. You can see them where coal mines rip off mountaintops in West Virginia and in oil-slick footprints on the beaches of the Gulf of Mexico. A square mile of industrial soybeans. A diamond mine in Rwanda. A closet stuffed with clothes. Windigo footprints all, they are the tracks of insatiable consumption. So many have been bitten. You can see them walking the malls, eying your farm for a housing development, running for congress."[7]

Ray asks, who protects climate refugees? Or Indigenous peoples whose very livelihood is tied to the land? "Will the climate-anxious recognize their role in displacing people from around the globe? Will they be able to see their own fates tied to the fates of the dispossessed? Or will they hoard resources, limit the rights of the most affected and seek to save only their own, deluded that this xenophobic strategy will save them? How can we make sure that climate anxiety is harnessed for climate *justice*?"[8] As lands burn, floods and hurricanes spur across landmass, earthquakes shift us palpably, the tectonic plates will continue to shift to the groove of their own intuition. We are at the behest of this Earth, not the other way around, and we are running out of time to listen to her and to meet her demands.

North Americans, Westerners, people who live on colonized lands as settlers, as I do and I have in three nations, three vast continents, owe everything to these lands and to the people whose lands are still unceded. I was born on Anishinaabe land, raised on the land of the Gadigal people in the Eora nation, and now live on Tongva land. This Earth, and the custodians of it, the people who have cared for it, who knew how to tend to it, to hear her speak through the weave of spiders, through the call of the hawk, through the pitch of the wind, they knew, and have always known, more than us. We owe them not just their land, we owe them our lives. We owe them for all that we've taken and continue to take without concern, and we show our respect for what they have given us, and what we have stolen, by being and becoming ardent warriors and defenders of this Earth.

"In the indigenous worldview, a healthy landscape is understood to be whole and generous enough to be able to sustain its partners. It engages land not as a machine but as a community of respected non-human persons to whom we have a responsibility," writes Kimmerer. "Biocultural restoration raises the bar for environmental quality of the reference ecosystem, so that as we care for the land, it can once again care for us."[9] In Maxine Bédat's *Unraveled*, she discusses the effects of industrial farming on the soil in an enlightening interview with Carl, an organic farmer in the United States who tells her, "Not using synthetic chemicals—that is, organic—can help to improve soil health and thereby prevent the need for fertilizers and pesticides to begin with. Within five or six years of being organic, you can start seeing changes in the soil—its crumble, ability to retain moisture, nutrient level, even the smell are all different."[10] Yet there are so many farmers who use the herbicides and pesticides because that's the dynamic of commercial farming; it's a naive belief that you can be in control of ecology, of harvest, and that it won't have grave impacts and repercussions. As USDA soil scientist researcher Rick Haney puts it, "We are essentially destroying the functionality of soil, so that you have to feed it more and more synthetic fertilizers just to keep growing this crop."[11] It's the cycle of capitalism to want more, it's *windigo* thinking, and this—our need for things is the very thing that needs to change. But before that, we must come back to the land. We must rematriate her, respect her and find ways to be in true communion with her.

"Among the three human obligations, the third is paramount," writes Bri. Maya Tiwari in *The Path of Practice* about the declaration of the Vedas. "Reverence to the Divine, reverence to Nature—to the earth, river, wind, fire, and space, to the animals, plants, and every blade of grass, every speck of dust."[12] I carry this with me as I look toward my life now, thinking about how my life changed when I found humility, when I found sanctuary in the land. I came back to her through the use of sacred medicines, and she has reminded me of her compass in all of us, how she is always there when we need her. I think of the power of sacred medicines and that to sit with plants, like Grandmother ayahuasca, mushrooms, guided by elders and the traditional custodians of those medicines is a privilege. I am humbled when I think of how the Huni Kuin, the indigenous peoples of Peru and Brazil, have received downloads from Grandmother to share her sacred wisdom with the world. There is so much intangible wisdom that we've blocked ourselves off to so much as a species. There's so much we refuse to see, or account for, but the plants around us, the Earth, I believe, is asking for our full attention.

As a Muslim, I have always been prepared for death, have had a specific awareness of it. This life, this material plane, is just one part of existence. Empires collapse and we die. That is what will happen to me, and it's what will happen to you. So, what do we do in the interim? In this space where we have found each other, found this ground, found this universe where we are all alive, facing *these* realities. We have the power to change, we have the power to *make change*. We have to understand that we are interconnected, and that global capitalism is an insidious fraud that compels you to believe you don't have any responsibility for what you buy. That extracting from people and the planet has no impact on you, but it does. You are complicit, we all are. It's not about guilt, it's about truth. It's about healing in a truthful way.

In the question of who is wellness for, I've come to understand that wellness isn't for anyone if it isn't for everyone. Otherwise it's a paradox.

As I slow down and sit with my own mistakes, I find myself full of compassion for others. I stopped needing to be right a little while ago, and that's been liberating: to accept that I'm faulty, as we all are. If we stopped targeting others, if we stopped severing relationships and connections, only to replicate the brashness with how we have been taught to treat the land, we could nurture a shared reliance on each other. This is what justice feels like for me.

CHAPTER 14

ON DEGROWTH

On February 13, 1960, during the Algerian War, France conducted its first nuclear test, known as the Gerboise Bleue, in the Sahara Desert. General Pierre Marie Gallois, a French air force brigadier general and geopolitician, was instrumental in the constitution of the French nuclear arsenal, as well as the Gerboise Bleue—an atomic bomb that was *four times the strength of Hiroshima*—and was given the nickname of *père de la bombe A* ("father of the A-bomb").[1]

A total of seventeen tests were carried out, four of them atmospheric detonations, and thirteen were underground, and it is said that the radiation caused a reduction in livestock and biodiversity as well as the vanishing of certain migratory birds and reptiles. The tests even led to the movement of sand dunes.[2] "These nuclear tests need to be seen in the context of a cruel and inhuman colonial experience that was synonymous with expropriation, genocide, racism and pauperisation," explains Hamza Hamouchene, cofounder of Algeria Solidarity Campaign and Environmental Justice North Africa, the "nuclear waste remains in the region with the French state refusing to take action to—literally—clean up its (radioactive) mess."[3] The International Campaign to Abolish Nuclear Weapons (ICAN) called on the French government to take responsibility for the long-term damage that it has caused. In a report last year, the

Nobel Peace Prize–winning group highlighted that "the majority of the waste is in the open air, without any security, and accessible by the population, creating a high level of sanitary and environmental insecurity."[4] This is how colonial forces have not only contributed *consistently* to the unsafety of those they colonized, but the hubris that these nations carry without any remorse for their actions is a telling sign of the West's incapacity for accountability.

I've realized time and time again while writing this book that there is a deep correlation between trauma and society, and how trauma exists and lives in society, within the most mundane social dynamics. It seeps into our literature, our art, and even in how we are governed. We are a species that has evolved to have little to no regard for each other, and as colonial forces prance around speaking about the indivisible superiority given their supposed democracy, they deny a history of inhumane decisions that subjugated more than half the global population. I think back to the words of W.E.B. Du Bois almost a hundred years ago: "Whiteness is the ownership of the earth forever and ever."[5] As white supremacy threatens to plummet the entire world into ecocide, is it not the responsibility of the empires that have built myths of self-importance, that have conquered land, wildlife, and peoples, is it not now their time to appropriately step up and protect the planet? Or do "superior" people not care about anybody else? Is that how we see power? Is that what we're all working toward? How long must the charade of white supremacy exist? All he brought us was an apocalypse.

"If there is any one thing that global warming has made perfectly clear," Amitav Ghosh wrote in 2016, "it is that to think about the world as it is amounts to a formula for collective suicide."[6] With Covid, this is even more apparent: "The societies most geared toward individual profit, and most worshipful of economic expansion, have proved least capable of saving themselves," Ben Ehrenreich wrote in *The New Republic* during the pandemic. "Decades of almost unbroken GDP growth have piled up riches in a few gated compounds while leaving the vast majority of Americans poorer and more vulnerable to illness, imprisonment, homelessness, and the ghastly futures that we know all too well await us. Covid has charted a precise map of its variegated terrain, of who gets to live and who gets pushed out to die. The same map applies to the climate crisis, too.[7] The most recent Emissions Gap Report from the United Nations Environment Program (UNEP)" encouraged the richest 1 percent to reduce "'their current emissions by at least a factor of 30,' which would allow the poorest 50 percent of the planet's population to *increase* their per capita emissions 'by around three times their current levels.'" "For the latter," Ehrenreich writes, "a threefold jump in consumption is the difference between constant want and a life of basic dignity. Billionaires who drop to 1/30th of their fortunes are still multimillionaires."[8]

In Anand Giridharadas's *Winners Take All*, he explains that neoliberalism, in the framing of the anthropologist David Harvey, is "a theory of political economic practices that proposes that human well-being can best be advanced by liberating individual entrepreneurial freedoms and skills within an institutional framework characterized by strong private property rights, free markets, and free trade."[9] This is all good in theory, yet the logistical question remains, *always*: Who gets to participate in this economy? And who has to be the labor force for the things we want? My mother would often chastise, "Money doesn't grow on trees," as I've grown older, it's clarified, because neither does labor. Every Amazon box delivered, every sneaker made, every crop that harvests the organic kale, corn and sweet potato—someone's labor went into that, so we could have, and attain, the lives we get to live. I keep thinking of President Charles de Gaulle, who, after detonating Gerboise Bleue in the middle of the Sahara, during a war, without any concern for the ecology or people of the desert, ecstatically exclaimed: "*Hourra pour la France! Depuis ce matin, elle est plus forte et plus fière.*" (Hurray for France! Since this morning, she is stronger and prouder.) How do we begin to move with intentionality, when the examples of that in our society are so limited and rare? Just a few days after the torrid bomb erupted, Morocco (which has dominion over the section of the Sahara where the bomb was detonated) withdrew its ambassador from Paris. Other African nations expressed their disappointment with France's decision to test nuclear weapons in the Sahara, citing fears of radioactive fallout and the safety of their citizens. It's evident that leaders like de Gaulle show an almost pathological disregard for those (and what) he (and the legion of other Western leaders) felt entitled to. As Silvia Federici writes so acutely in *Enduring Western Civilization:* "Indeed, many of our political rights were wrenched into existence against the resistance of the most typical 'Westerners.' The 'Western Civilization' 'legacy' metaphor also hides the role European and non-European workers (both were considered outside the pale of 'civilization') have played in building the wealth and culture of Europe and America. Typically, credit for technological development is laid at the doorstep of Greek Rationalism or is presented as the logical unfolding of a Promethean inner 'Western' predisposition; rarely is it asked 'Who built the factories?'"[10]

In 2002, in a piece about the prolific Kashmiri poet Agha Shahid Ali, Ghosh, a Bengali-Indian, quotes a line from Ali's poem

Farewell: "Your history gets in the way of my memory."[11] The poem itself reads as a plaintive letter from a Kashmiri pandit about Kashmiriyat—the centuries-old syncretic identity of Kashmir, which merged Hindu-Muslim culture, festivals, language, cuisine, and clothing—and came under assault in the early 1990s. There's a metaphor lodged in the reality of Kashmir, as it is a huge point of contention between India and Pakistan to this day. Potentially because it remains as an artifact of a bridge between two worlds. A pre-colonial syncretic sensibility.

Instead, now we exist primarily in separation from each other and extraction of each other. According to Bitch Media, "the global Ayurvedic market alone is expected to be valued at $14.9 billion by 2026. Indian exports (especially herbs, spices, and medicines) made up a sizeable amount of this market in 2019, and Prime Minister Narendra Modi has credited the Covid-19 pandemic with a 45 percent increase in exports of Ayurvedic products between September 2019 and September 2020,"[12] products produced by the very farmers protesting. And, though nearly 60 percent of the Indian population works in either agriculture or an allied sector, Arundhati Roy writes, "Indians are too poor to buy the food their country produces." What Winston Churchill said of Palestinians comes to mind here, as it's applicable to every plight for self-determination against colonial power: "I do not agree that the dog in a manger has the final right to the manger, even though he may have lain there for a long time. I do not admit that right. I do not admit, for instance, that a great wrong has been done to the Red Indians in America, or the Black people of Australia. I do not admit that a wrong has been done to those people by the fact that a stronger race, a higher grade race, a more worldly-wise race, to put it that way, has come in and taken their place."[13] Ah yes, I wonder if "worldly-wise" here is code for exhausting hubris. The aggressor makes the standard, and then everyone has to play by their violent rules.

There are many adept (and funny) memes that reckon with these ideas far more eloquently than I, but I shall attempt to point out the obvious: Do we ever stop and think about how our limited imaginations brought us *here*? How we literally could have devised anything on Earth and instead we decided to create scarcity, hatred, racism, and poverty? It's not as if these things can't be effectively eradicated—it's as if people are uncomfortable to change, to believe *in more*. Isn't racism largely fueled by a fear of scarcity? That what you deem as yours and only yours will be taken by some unknown stranger?

But what if we committed to looking at the facts? "To say that the United States is a colonialist settler-state is not to make an accusation but rather to face historical reality," writes Roxanne Dunbar-Ortiz.[14] Healing personal traumas is similar to healing societal wounds. Behaviors are redeemable, but humans, it seems, are immensely lazy. It's easier to just *say* you're this and that instead of actually investing the time or energy to *be* this or that. Many of us have been compromised so many times that I imagine it has made us impenetrable. The hardness becomes a shield, but now, after many years of denying your feelings, you can no longer access a sincere emotion. Everything is rote or a reaction. Nothing feels grounded or like your own. This desensitization from yourself and your integrity continues the cog. It's our collective confusion that further pulls dishonesty. The Red Nation writes in *The Red Deal*, "The cunning politics of injury has come to define our era, where state and corporate perpetrators of violence and genocide are asked to—or position themselves—as purveyors of 'justice' while entirely sidestepping the question of colonialism and imperialism."[15] The Navajo Nation is *still* one of the largest resource colonies for the United States—providing natural gas and supplying energy through coal, and yet many of its citizens still live without clean water or electricity.

◎ ◎ ◎

There is a phenomenon known as postapocalyptic stress syndrome, which is the result of an intense shock from which a culture never recovers. "The Europe that came out of the Black Death was not the same as the Europe that went in," Dr. Larry Gross tells Tyson Yunkaporta in *Sand Talk: How Indigenous Thinking Can Save the World.* We are facing an unprecedented time, and as Yunkaporta tells us, it makes sense for us to act fast so we don't obliterate into extinction. "We need to start working with the land, rather than against it," Yunkaporta warns us. "Our communities need to share knowledge with one another while maintaining their own unique systems grounded in the diverse landscapes they care for."[16]

Copyright is an American invention, my father explained to me over the years, and by introducing Vandana Shiva's work to me, he helped me begin to understand how biopiracy* has played an important part in the conversation about wellness. In *Earth Democracy,* Shiva explains how important the seed is. "The highest sin is to allow the seed to go extinct. In my region in the Himalayas, during the Gurkha war, there was starvation but in not a single hut were

* According to *Merriam-Webster, biopiracy* is defined as: "The commercial development of biological compounds or genetic sequences by a technologically advanced country or organization without obtaining consent from or providing fair compensation to the peoples or nations in whose territory the materials were discovered."

"Biopiracy," *Merriam-Webster,* https://www.merriam-webster.com/dictionary/biopiracy.

the seeds eaten. The seeds had been left untouched for future generations."[17] For Shiva, it's the lack of respect that colonizers have for the land that has contributed to their craven extraction of everything. When people can't be in right relationship with the Earth, if they can't honor her and see her unique intelligence, how can they see humanity in others? "Everyone is panicking," Shiva said recently in an interview with *MOLD* magazine,[18] "But all you have to do is turn to the seed to learn how to address these mega problems: self-organization—becoming part of a living and generous Earth—and learning the economy of abundance rather than subscribing to the economics of scarcity."[19]

The charade of masculinity and the banality of domination is the undercurrent of our entire species. The patriarchy disdains what it believes is extractable, and thus weaker, because in its own conception there is only room for the aggressor and submitter. But isn't masculinity itself an imagination that is propagated through performance and bravado? It's not mysterious, there's no liminality—that's why masculinity is so boring. It's a construction that remains unchallenged by most, as if it's just fact. But we are not machines, we can choose our own destinies.

Shiva speaks of how when the British took control of India, they considered one third to half of the total area of Bengal Province as unnecessary wasteland because it wasn't farmland—it was forest—and therefore yielded no revenue. I think of stories of the forest that I grew up with as a child, the lush groves that are home to the beautiful and sublime Bengal tiger. When I was younger, I would pretend I was there, a *Rajkumari* in the wilderness. But my home was stolen by a thief, a thief who deemed the forests of my unknown youth wasteland. By people who believe that only continued dominance can yield results.

Shiva does the due diligence of explaining the Indian concept of *bhaichara* meaning a custom (*achara*) of the brother (*bhai*)) that was common before British colonization. The basic understanding of *bhaichara* was that a family would only pay taxes on what was earned by them as a family and instead meet them halfway. "Colonialism reversed this disbursement ratio, with Britain leaving only 10 percent for local infrastructure to sustain the people and taking 90 percent to run the empire,"[20] writes Shiva.

Many cultures were *not* built on aggression or on conquest and war, and it's time we see value in other ways of being. "Humanity would not exist without caretakers. But caretaking is labor. It takes work to plant crops. It takes work to hunt. It takes work to raise children. It takes work to clean homes. It takes work to break down a buffalo. It takes work to learn the properties of traditional medicines,"[21] writes the Red Nation. It's true, the cultural attributes of the colonized were rarely seen as valuable. But land and resources weren't the only things that were taken. Leanne Betasamosake Simpson states, "Sexual violence is an effective tool of conquest because of the overwhelming damage it inflicts upon families, lasting for generations and instilling shame and humiliation that discourages any efforts to resist." In North America, sexual violence was also way to disconnect Native people "from their source of strength—their families, cultures, traditions of resistance, and the land itself." Throughout the boarding school era, "Native children were frequently subjected to sexual violence and physical abuse by teachers and school administrators, including Christian priests."[22] Then if you were a Native woman, you were constantly subjected to the fraudulence of the American healthcare system through a loss of complete agency "via forced sterilizations, lack of access to equitable prenatal and postnatal care, poor reproductive health services, poverty, assimilation, exotification, and objectification."[23] This violence is deep. But as Chief Seattle of the Suquamish and Duwamish nations reminds us: "All things are connected."

◎ ◎ ◎

"The fashion industry more broadly is believed to contribute between 4 and 8.1 percent of the world's total carbon footprint," writes Maxine Bédat, adding: "those in the environmental community tend to think of our climate impact as things that take place within our own borders. But this is deceiving. As we see with our clothing, our carbon footprint extends far beyond our borders and must be taken into account when developing carbon reduction policies."[24] Bédat explains the breakdown, how most garment workers make 0.4 to 0.5 percent of the retail costs of the clothes they produce, how the atrocious and inhumane working conditions, with an addition of limited pay, generates and sustains a loop of poverty. A study in Bangladesh found that a vast majority of garment workers suffer from malnutrition, 77 percent are anemic, and many are underweight. "It's undeniably positive that unionized American workers have raised their voices for better conditions, but the lack of awareness about global labor pits US workers not only against themselves, but against the rest of the world in a cruel Robin Hood trade," adds Bédat. "Without international labor standards included in our trade deals or enforced through our own domestic laws, the more 'costly' protections workers in the United States demand (whether it's higher wages or benefits like health insurance and retirement plans), the more likely companies will be to seek labor that's less costly, from workers across the border and overseas."[25] We have to begin to see this holistically, as all things are connected. We have to act with that in mind.

"The seed is a circular economy and that is the economy of abundance—it is the economy of sustainability; it is the economy of renewal; it is the economy of never getting exhausted,"[26] Shiva explains. I wonder if the extractive ways in which Indian farming has evolved have created an even more harrowing situation. Against the backdrop of the Farmer's Strike, the three farm bills were light-years away from the concept of *bhaichara*. Unsurprisingly, between 1995 and 2014, a reported 296,438 Indian farmers have committed suicide.[27] We've come so far from our roots in an attempt to become more civilized, we've lost our very humanity. "In Eastern cultures," Shiva laments, "we've always seen eating as an act that embodies our relationship with the earth—if it is a conscious act. A person who's not a farmer can still be a co-producer with the farmer through the consciousness of what the farmer is doing and gratitude for farming with care." We need to regain this esoteric connection we once had with each other. We need to reinstall the spirit of dependency, of care.

Back to Dadabhai Naoroji's drain theory, three journalists for *Al Jazeera* recently wrote about how drain theory continues via poor wages, as workers from the Global South get, on average, one-fifth the level of Northern wages, meaning "that for every unit of embodied labour and resources that the South imports from the North, they have to export many more units to pay for it."[28] Incurring, again, a cycle of poverty for complete extraction. "Economists Samir Amin and Arghiri Emmanuel described this as a 'hidden transfer of value' from the South, which sustains high levels of income and consumption in the North. The drain takes place subtly and almost invisibly, without the overt violence of colonial occupation and therefore without provoking protest and moral outrage."[29] This has resulted in an estimated 152 trillion dollars from the Global South since 1960. We perpetuate a farce when we say "developed" versus "developing" without the context of how the West creates debt in order to cajole and suppress, and that control over the market—and whose dollar has more power, or again whose value is more valued—are all invisible gears that are churning a very old story.

This is where the stories of the farmers and garment workers intersect; it's their labor we are taking, valuing capital more than bodies. "Rich countries have a monopoly on decision-making in the World Bank and IMF, they hold most of the bargaining power in the World Trade Organization, they use their power as creditors to dictate economic policy in debtor nations, and they control 97 percent of the world's patents. Northern states and corporations leverage this power to cheapen the prices of labor and resources in the global South, which allows them to achieve a net appropriation through trade."[30] It shouldn't come as a surprise that the laws are corrupt and so are the regulators, but we have to see that all of this is connected and that the tradeoff of destroying the Earth for the sake of profit is also no longer an option. We must insist on change at every level. It was a grassroots climate movement in the United States, Bédat explains, whose protest of pollution led to the Clean Water Act in 1972 and the Environmental Protection Agency (EPA). This "created laws and an enforcement mechanism to limit pollution from the textile and energy industries."[31] Yet, as Bédat tells us, "when American companies moved their production to China (and elsewhere), these protections were disregarded."[32] Again, because human life, if it is not American, Western, or white, is seen as disposable, and this is apparent in every facet of our modern lives.

Giorgos Kallis, in his book *Degrowth*, contends that this "social metabolism" of our current age has led to "the growth paradigm: the idea that perpetual economic growth is natural, necessary and desirable."[33] This concept of "social metabolism" became the central tenet of the geopolitical world order from the 1950s onward, resulting in the "Cold War and the arms race, the end of colonialism and its indirect continuation under the guise of 'development,' and the failure of socialist projects for equality."[34] According to Kallis, no economic growth can be ecologically sustainable, and as we are facing planetary ecological breakdown, our time is limited to make the necessary changes to prevent complete catastrophe. I am haunted by the image Bédat creates of a black river in Bangladesh: "Three of Bangladesh's rivers are now 'biologically dead,' supporting no forms of life."[35] This is an image of catastrophe, of death, of complete and total ecological collapse.

Degrowth evolved as a critique of economic growth and was originally termed by Austrian French social philosopher André Gorz in 1972, who argued that global environmental balance, which is predicated upon non-growth (or "degrowth"), is not compatible with the capitalist system, which requires "accumulation for the sake of accumulation." The defense of growth by nation-states is then used to argue that exponential growth is needed to overcome poverty and to create jobs. "This is bourgeois ideology in the sense that capitalism relies upon and produces the *artificial* scarcity to which we're subjected,"[36] writes Joseph Hickel, an economic anthropologist.

Human relations and enterprise have been around for millennia. I think of a young Prophet Mohammed working in the busy cosmopolitanism of Mecca for his first wife, Khadija, a businesswoman, divorcée, and fifteen years younger than Mohammed. I also think of the wild majesty of the grand bazaars in Istanbul, Casablanca, the marketplace of the Silk Road—where goods were exchanged and valued, and which existed long before the introduction of capitalism. "Indeed, what we today understand as 'economic' activities were once embedded in social institutions in pre-capitalist societies like rituals, kinship networks, and state or religious mechanisms of redistribution. Market activities were subordinate to politics and values."[37] Now there's no focus on the value of human life, and therefore upholding human *values* has dwindled. Capitalism has merged with the state and has become deeply embedded in it. Exploitation comes easy when there's no barometer of what is moral, and there is no true objective regulator when all your courts can be bought.

The concept of degrowth asks society to head toward a "radical and egalitarian social transformation" by literally degrowing. Buying less, consuming less, using only what you need. But it's an ideology for the privileged, not for the poor or homeless. We have to understand that there is a cost for our consumption, and yet we blame carbon emissions on the developing world, on the poor, without considering what created this division. "The wasteful and environmentally unsustainable consumption patterns of the working class are not produced by 'personal' choice but are system-induced. Every day, millions of workers in the U.S. commute to work in single occupant vehicles not because we 'choose' to drive. It's because public transportation is so unreliable, jobs in the labor market are so unstable and temporary that few workers are actually able to live close to work," writes Collin Chambers.[38] Then of course the planned obsolescence of electronics—take Apple changing their chargers every few years so you're forced to buy a new one—is a predatory way that capitalism encroaches upon you, forcing you to succumb to unethical bind. But what if we built things to last? As it was for our parents? Why is "developed" synonymous to "surplus"? This waste that we've created is a deeply immoral pact we've made at the expense of the poor and of the Earth. Through actualizing degrowth, Kallis explains that it is possible to sustain well-being and improve living and ecological conditions in "an economy that unavoidably will contract."[39] This transformation will help us braid the economy back into society to ensure that nobody faces scarcity, which will inevitably mean creating limitations on how much we consume. Hickel advocates for "unilateral decolonization," calling for the South to enforce degrowth in the North "by refusing to be used as a supplier of cheap labor and raw materials for Northern consumption."[40] Ending this exploitative relationship, Hickel tells us, would require "Northern countries to either pay more for resource and labor imports from the South, or otherwise to rely on their own resources and labor. Both options would be more expensive,"[41] so Northern countries would naturally have to consume less.

In the late 1980s in the state of Gujarat, India, a network known as the Honey Bee Knowledge was engineered to create safety and precautions against "knowledge asymmetry," which is what occurs "when the people who provide knowledge do not benefit from the gathering and organizing of that knowledge."[42] The system is based on the metaphor of a honeybee that collects pollen from the flowers without damaging or exploiting, instead continuing the cycle of life, giving abundantly, openly.

In the Mpumalanga province in South Africa, eighty traditional healers whose health and medical training came from traditional apprenticeships, created a biocultural knowledge commons to both archive their ancient knowledge and better serve the health needs of the people actually living in their province. "An outcome of virtuous relationships with the land, the plants and animals," they explain "is not property to be bought and sold. It is simultaneously cultural and spiritual and its movement and application promotes a kind of virtuous cohesiveness." My favorite story is one of Paulo Wangoola, a Ugandan intellectual and civil society activist, who, after working in various parts of Africa and abroad for over two decades, returned home to the Kingdom of Busoga with this message: *Take heed, this ancient knowledge cannot be lost.* "You sent me out," he told his elders, "to gain Western knowledge and to work in the structures and organisations of the Western world. I have been to their universities, have worked with their governments, have created Western style organisations here in Africa and now I have come home to share what I have learned." His message was simple: "The children of Busoga Kingdom, the children of Afrika will never realize our full potential as people in our communities and as contributors to the global treasury of knowledge if we continue to depend wholly on the content and ways of knowledge of the European peoples. Our way forward must be linked to the recovery, replenishment and revitalization of our thousands of years old Indigenous knowledge."[43]

There is wisdom in trusting your instinct. Wangoola went on to establish the Mpambo Afrikan Multiversity,[44] a higher education research institute for the revitalization and promotion of mother-tongue Afrikan scholarship, dedicated to protect the knowledge keepers, where he is currently president.

Degrowth, as Kallis explains, is inevitable, but it's the path that we choose forward that is most important—one that integrates the vastness of colonial rule and how insidious these structures have been. It's as if colonizers wanted our complete subservience for time immemorial. These days, it feels like there's no escaping the root of white supremacy; a people that are *this* hateful must hate themselves so much. I'm not being facetious. As a rational person, and as a writer and thinker, to me the construction of whiteness is baffling in many ways, like masculinity before it. Capitalism next to it is an obnoxiously cruel system, and yet it continues. But there's a shift happening, and as the Earth speaks louder and louder, there will be fewer chances to look away.

In *Suicide,* Émile Durkheim writes that "people who are more tightly bound by ties of family, religion, and local community have lower rates of suicide. But when people escape from the constraints of community they live in a world of 'anomie' or normlessness, and their rate of suicide goes up."[45] What if all people were held? What if we all felt wanted? What if some families weren't just surviving through trauma? "In the United States, suicide is the second leading cause of death for people between the ages of 15–35, a reality that closely parallels statistics for the rest of the world, where suicide is the second leading cause of death for people aged 15–24," writes the Red Nation.[46]

As a person living in those demographics, this doesn't surprise me. It's a sense of fatalism that is evident as we hurtle toward darker and more difficult times. I've never understood nihilism, and its abjectness seems like just another facet of whiteness and the mythology that sustains it. Because, frankly, it's boring that a people who divided, conquered, slaughtered, and plummeted the world on fire has the arrogance to think life has no meaning. It had meaning for the people that were forcibly removed from their cultures, their homelands, their families. It's been interesting to write this book during a pandemic, to consider what wellness is *now,* because all the answers point us to a time before the construction of these horrifying systems.

In one of my favorite Toni Morrison talks, she expresses the power of goodness. "Evil is silly, it may be horrible, but at the same time it's not a compelling idea. It's predictable. It needs all this costume to get anybody's attention. But the opposite, which is survival, blossoming, endurance—those things are more compelling intellectually, if not spiritually."[47] I spit in the face of nihilism. My ancestors survived an apocalypse, so many of ours did, and yet here I am, writing to you. To me, real human conviction, the courage to be good—even when all you've faced is bad—is magic. It's to learn how to alchemize, to make shit into gold. How else can I explain being here? Other than that I tapped into an infiniteness that had been quashed and shattered, but I never surrendered to the pain and the anger of that hurt. To me, goodness is far more interesting, because *it is* harder. Society should change, and it should evolve even if it's harder work. If governments are failing us (which they are), we have an option to completely rewrite the script so that nobody faces scarcity.

I've been part of the New York chapter of Degrowth for a year and a half now. Every meeting I'm humbled by the work, I'm moved by the desire and grace to want to change things. I don't have the answers, but I'm willing to learn how to move forward. I know in order to participate in the future I want to live in, it's my responsibility to act, to consider, to change for good. In a meeting with Kallis at the end of 2021, he told us to shift toward an "environmentalism of the poor" mentality, a method that encourages and challenges the elite to change, first. The elite are the most resourced and thus have a duty and responsibility to act. We have to learn how to engage in society equally, for a collective effort and benefit for all. More of us need to stop pipelines from being built and encourage governments and their leaders to dissolve NATO, for example. We have the power—and if you have a lot of money . . . then you really do. Rich people have the ability to completely change the world, they just choose not to. It's unconscionable and it makes me sad.

Kallis told us that if you have a major platform it's imperative now you use your voice to rally against corporatization and encourage true sustainable methods of being with the land and each other. In a climate apocalypse, what else do we have but this? Building toward one another, without the apparatus of fictious power. Degrowth might be our only option.

CHAPTER 15

ON HEALING OUR WOUNDS WITH THE FEMININE

Friendships with women mean a lot to me. Historically this has always been the case. As someone who went to a public all-girls' school, someone who has been queer since she was a teenager, I cultivated a certain kind of love of women and femme folks that was constantly being triggered and tested by my mother. I wanted to believe I could have healthy relationships with women and girls, but the person I hated the most was a woman, and her ideas of women were plagued by a simultaneous prolonged fascination and deep hatred that she had seeped into my bloodstream, haunting me with her bromides. Because no one in my family was allowed a social life, my mother made up stories about my friends. Sometimes they were funny, if she was in a good mood. With her vicious, sly, and rank type of humor, she'd make a raucous joke that would momentarily make you believe she was on your side. If she was in a bad mood, she would call my friends sluts and whores and forbid me to even talk to them at school. Still, my relationships with women in my life have also been small miracles, albeit also some of the most difficult relationships I've had.

Over the years, I've conditioned myself to be a people pleaser; when I have stopped playing that role in people's lives, ten out of ten times I'm accused of being selfish, of being mislabeled a narcissist, of caring only for myself in the aftermath. These are words I'm familiar with. Choosing anything for myself as a child warranted being lambasted as selfish by my mother. I believed this was true until I began to live in the world and saw how other young girls lived, how they thrived, how they were treated by their mothers or families. I realized I had been duped and denied the foundational respect a child deserves for just being.

Everyone in my family has a martyr complex, so passive aggression was rampant. If you weren't martyring your happiness for someone else, were you even a good person? This was something that lingered like a cruel sentiment in interactions with my family. We were all so depressed and incapable of knowing how to pull the other out because we were all implicated in the violence my mother spawned.

When you grow up in abuse, everything becomes part of the grooming process. Conditioning is so important, and yet we rarely fully factor it in. When Gabor Maté writes that "many people unwittingly spend their entire lives as if under the gaze of a powerful and judgmental examiner whom they must please at all costs,"[1] what he's identifying is abusive parenting and family dynamics that become mantras of self-policing. The mind grips on to the grooves of the language used by our parents like they are lullabies, good or bad, nightmare or dream. Nostalgia brings us back to these formative years like a lighthouse. I deserved the life I had been given, *I told myself,* and by witnessing the actions of my sister and my father, I believed that my duty was to give my life over to the family. Even if I was bad, by the light of my sister and father, I knew what goodness was. God felt close in those days, too, because I needed a narrative to sustain myself. I wondered if all that was happening was a test of character. I wanted to be a good girl, I wanted to be a good person for my family and for God.

Still, I fail, every day. As my female friendships in my teen years dwindled, I found I was the only one in my friend group who craved depth all the time. A lot of people call me intense, possibly because I can't have a banal conversation, I can't perform. I don't understand social niceties, and I refuse to. I'm notoriously blunt, and this, alongside my kindness, can't be understood, so I get misrepresented and mischaracterized a lot. Especially by other women. I'm sure plenty of people call me crazy. These days that brings humor more than it does pain, because the way we engage with each other is so scripted, so pathetically predictable. We are keys to each other's liberation—why do we keep forgetting that?

As I get older, I've noticed I care less and less about what others think of me. I've found this phenomenal dance with myself, a cadence and grit, a movement of understanding that these depths I seek are available, and they are portals to get closer to myself. This has helped me ground myself in my own inner knowing and self. I guess I care less about what others think of me now because I've gained self-respect through this journey of healing.

But this feeling of being a social pariah, especially in relationships to other women, has developed into wanting to understand my failings. If I was the problem, I wanted to know why. I didn't have the bravado that some have; instead I had deep insecurity and a heart so big and wanting to be seen. But didn't understand why it kept getting rejected or misunderstood.

In my late teens, just before I graduated high school, most of my closest girlfriends entered relationships, and I remained single.

When I went to university, I was young and messy, a new flavor I'd never been allowed to try because I suddenly had freedom. I was at a loss with myself, and no matter what I did, I was unlikable. I began to disassociate again. I would lose most of my friendships mainly because I was a loner and a nonconformist. While I still craved deep friendship, the women closest to me rarely could or wanted to keep up with me. I was always the disposable one, and I got thrown out and around a lot. I was rejected for being who I am, for moving as myself. So this made me question who I was constantly, and thus made me feel more insecure, until I stopped. Until one day I just went, *Nope. I'm not participating in this charade anymore. I refuse to metabolize other people's insecurities of who I am.*

Another thing about abuse is that when you haven't articulated its parameters to yourself or to others around you, people assume you grew up in a normal environment like they did. Even if they also didn't, they will still project normalcy onto you, and if you go off-script, my oh my are they quick to critique, torment, and laugh at you. Though many people have had shitty childhoods, there is a spectrum. It took me a long time to accept that a person can only ever meet you how deeply they meet themselves. As I grew older, I began to see how women treated me different from other women, and how often I was relegated into a role of silence or the butt of the joke. I saw how quickly, easily, other women relied on talking down to me or making me feel bad by being overly, exhaustively critical, wielding shitty comments that triggered me, once again relegating me to the role of the hurt, confused, and over-sensitive type. If I reacted—which I often did—I was castigated even more. I wasn't liked, I assume, because I've never been able to fall into place. It was my fault I couldn't get in on the joke, but I wasn't sure why *so many women* had to interact with me like this, how I perpetually felt I could only be liked if I was small against someone else. If I was me, the full me, I was too big.

When I was younger there wasn't the societal awareness we now have of trauma. Culturally things were so different, emotions were corny, and as a person with many, I realized I was better at hiding and serving. Staying quiet is a trauma response. People liked me better, too, when I was this way. When I gave more, because I took my place in the social hierarchy that was governed for me. There were a couple of years in my twenties when I thought I could get into line and just stop the restlessness by being how others wanted me to be, servile. I told myself I was worthless and that what I wanted was so impossible that I settled for people who petrified me into a perma-state of guilt. I had good friends but somehow found myself forever drained by most people. Still, I blamed it on myself, like I was the eternal problem. I funneled the frustration into my writing, thinking about all things macabre, but also, again and again, about friendships between women.

My former friend C was a beautiful woman in a way all women want to be. Her body had a dancer's grace, and she was quirky in a *real* cool girl way. We were good friends, and I had longed for her friendship as I often longed for friendships with women, especially with other women of color. Throughout our friendship, I fawned over her. Dripping with compliments and earnest observations, I quickly became a confidante, and I took the role seriously. I believed good girlfriends were truer than gold, but when one day C told me she felt jealous when people told me I was beautiful, I felt a bit off. I was the frumpy one in the friendship. I had forever been the frumpy one, incapable of self-actualizing into my body; it constantly felt like a thing outside of me. I assumed most people could read this on me, and that kept me even smaller.

My life had been forged without me for so long that by fashioning myself against the beauty of my friend, I had started to slowly feel . . . acceptable, as I had never felt myself to be. Being around her made me feel beautiful, because as I watched how she adorned herself, I started to do the same. She had known all this, known, at least on a surface level, how difficult it was for me to feel embodied. When she said those words to me, about her jealousy over a beauty I didn't know I had, instead of feeling the full force of the fusillade, I cowered and sympathized. Truthfully, I was glad she was being honest. I thought bluntness was what garnered good relationships with women. But what she was saying was really an observation without any desire to change it. What followed was a revolution on my end. A few years later, unsurprisingly, right as my career started to finally show signs of success, this same friend accused me of being a narcissist. Up until that point I had sought counsel from her about whether to post photos of myself that I liked on my Instagram account, which up until then had been a place of jokes or photos of other things, anything but myself. If I rarely posted a selfie—it was shadowed, hidden, I was still in so much shame. But there was also an itch that was brewing, and the itch told me that I needed to start liking myself. I liked the idea of Instagram as an archive and decided I wanted to start coming to terms and like this body of mine. Photos, I figured, might be a good way to confront my dysmorphia and dysphoria. It was slow, but I wanted to like myself.

I'm not perfect, nobody is, but it has been interesting to track and observe patterns I've had with some of my closest friends who are women. When they have failed in the past, I often thought it was my problem; when they guilted me, spoke about me behind my back, I secretly believed that maybe I was just a shitty human who had turned people away from me. But as I grew older, maturing into myself, I realized what actually was at play was that more women than not seem to hate other women. The oversaturation of misogyny becomes known as culture. I've found that when triggered by other women (for their beauty, style, grace, intellect, charm), instead of relaying that information in a positive way, we are encouraged to feel threatened. I was guilty of participating in this dynamic as well. This is not surprising when you see the information we are socialized with about other women. When I started therapy, a pattern began to emerge.

In *Women Who Run with the Wolves*, Estés writes, "When a woman has this ambivalent mother construct in her psyche, she may find herself giving in too easily; she may find herself afraid to take a stand, to demand respect, to assert her right to do it, learn it, live it in her own way."[2] There are many things therapy has offered me, but above all it has been a place where I've been able to cultivate self-confidence. I started to see how constant doubt about one's own value is often tied to parenting. If a parent constantly tells you what you feel is wrong, then of course that's gaslighting. These things are not just buzzwords of a zeitgeist, they actually condition you. The way I navigated the world was ruled by fear, searching for people who could protect me. I found that other women wanted my depth, but only for themselves. When I would start to need things or show any messiness—or, painfully, when my life was going superficially well—close female friendships would drop off, without a beat, blaming me for my sins, never once exacting their own. I'm not a person who hides, though I used to be. And I've noticed that every time I step closer into myself, even if it's a bumpy transition, people abandon ship. Over the years, I've believed it was a fault of mine, but these days what I think is that for all the charade of feminism, what our society really lacks is a real love for women—how dynamic we are and how beautiful we are in all that difference. We expect our lovers to love us, but so many of us lack real love for ourselves, love that isn't about clout, or beauty, or wealth—but a love that is deep and pure. Without conditionality.

So many of us fear that there can only be one, even if we're actively telling ourselves otherwise, that when someone poses a threat to our understanding of ourselves (and for many of us that sense of self is built on making others feel small), we are incapable of just letting that person be. Instead, we become jealous, vindictive, and hateful. Even if it's just in private. But this energy can be shifted into a profundity and awe for who we are and how much of this world we carry, despite existing in a world that wasn't designed with us in mind. Isn't it miraculous how we adapt, how we care, how we find ways to create and be? It's important for us to reckon with how we fan the patriarchal flame when we turn from and against one another.

When I first told my therapist that I didn't know how to love myself authentically, she told me to start respecting how hard I work at every facet of my life, not just at my professional work. I began to see she was right, and I realized, all these years as I had awaited validation from all these places—from friends, peers, institutions, lovers—when all along I could have invested in validating my own soul work. How time and time again, no matter how much I fell off the path, I still found my way back to myself. How incredible that after all these years I haven't abandoned myself. I'm still here, trying, attempting, open-hearted.

As imperfect people, I want to understand how we can have

more mercy for ourselves and others. That's why in the last few years I have considered myself an abolitionist. As an imperfect person, I'm thinking ahead, to the future. All of us will fail at things. We will try and we will fail. We will have good intentions yet somehow not be able to show up as our full selves. I have had to remember that, and every time I feel angry about my life, I am better served by looking at things more holistically. My shame of all the things that make me "not good enough" has kept me from living a good life, a life where I feel good about myself.

I wonder: if we could hold this fact close to our hearts, could we also understand how complex other people's lives are, and how people are going to hurt you just as you are going to hurt other people. If we could genuinely accept this, I believe we could shift in a major way. Relationships are important because in times like these, times of the apocalypse, all connection is *that much more* sacred, potent on an individual level. On a global, macro level, our relationships to and with each other determine our past, present, and future. Systems that have failed us and yet continue to reign—white supremacy, capitalism, patriarchy—are determined by, you guessed it, relationships. Yet how often do we truly examine this very fact? How connection to each other is at the core of who we are, it's why we turn to art, to songs and poetry, it's why we started documenting stories on cave walls and in books. Yet we've filled that space to be with each other with the endless pursuit of money and racism. In times like these, dark times, as we look toward the next stage of where we are going, we will have to prioritize connectivity more.

"The fear of loss of relationships is a key way in which patriarchy holds us back," writes Bethany Webster in *Discovering the Inner Mother.* "We keep our voices down, we exaggerate the truth, we manipulate to get our way, we stagnate our growth as a form of loyalty. These patterns are not a cause for self-judgment or self-blame. These are intergenerationally transmitted survival mechanisms for enduring misogyny, and a cause for self-compassion."[3] Women have been conditioned to hate women. We need to accept that as a fact, *for revolution.* Until we can unlearn and dislocate the misogyny inside of ourselves, we will continue to harbor insidious ideas about other women. And this will continue to affect relations with each other as well as the Earth.

I've seen this in myself, how I'm so quick to judge sometimes, but I wonder if it stems from how often I have felt people were quick to judge me. Maybe it's still in the pain of my teenage years of being left behind for boyfriends. Insecurities and projections (both mine and others'), and not speaking of them, along with jealousy, have created an enormous amount of tension with some women in my life. I have been saddened by how I felt I was intentionally misunderstood by them, by how the violence of my early life seemed to dictate the loneliness of my adult life, and yet not a lot of people could hold how the absence of a mother sits in your body. How cold it feels, and how visceral the pain is every goddamn day. So, narratives are drawn, assumptions are made about you when all you're trying to do every day is just survive. When we are unwilling to meet each other with genuine concern and care, what does that say about us? What does it say about wellness when so many women can't be well with each other?

For years, I would hold on to the righteous anger I felt from being blamed for things inaccurately or judged without compassion. But I began to see how little we want to give others the benefit of the doubt and how hungry we are for schadenfreude—especially in relations between women—and that began to scare me into changing. I didn't want to participate in this kind of relationship with anyone. I've seen this as a public person. I take note when women throw me under the bus, and I pay attention to who says what about me. I understand not everyone can like me, and not everyone wants to. I just wish it wasn't for superficial reasons that are coded in something else. Still, these days, I know women deserve *my* compassion. We need it *from each other*. We need it *for* each other. "Women's undiluted honesty is a huge threat to patriarchy," Webster writes.[4] There's a revolution here, but we have to learn not to stand in each other's way. We have to learn how to uphold the power of all women, and that upholding one's should never come at the cost of upholding another's.

In *You Belong*, Sebene Selassie explains that through the process of trying to belong, humans "tame our natural physical expression from an early age. We learn to police ourselves. Through our families, communities, and society at large, we absorb messages about what is acceptable and not acceptable in terms of our existence: how to move, how to speak, how to dress, and how to belong or not belong."[5] This we then use against others—we police ourselves and others against a weirdness that may or may not be present, against a beauty standard that's fake, against a body stereotype that's constructed. Yet all we want, all we really crave, is belonging. To ourselves and from each other. Looking to the future is understanding that the ways we've been programmed to be in society don't have to be that way anymore. It's clear to me every day the difference between committing to evolution and *not* is accepting that the journey is long.

We are a society that rests on the social behaviors we learned in our youth, with our families. When you decide you are not amenable, you accept your fate in a certain way. You accept your shitty, toxic behavior but refuse to accept others and theirs; you judge people for the very things you allow yourself or your friends to do; you cherry-pick alliances based on fake loyalty or camaraderie. How many times have I heard some lurid story about another woman, judged her for it, only to find out it was only partly true or the person who was telling me the story was friends with the person again? We can break from these patterns, but it takes collective awareness, accountability, and a desire to change.

We create false narratives to explain our feelings, sometimes believing them as intuition, but is treating another woman badly ever intuition? I've had to ask myself that question and change my own behavior, realizing how sneaky and mercurial feelings can be and how conditional behaviors and limiting beliefs not only affect us but the people in our lives as well. In moments where I've judged others, I was so absolutely convinced that I wasn't being harmful, but the reality is intention has nothing to do with harm, you can still be harmful when you didn't think you were, and I've had to accept that and take accountability. So, what then? Do I castigate myself for an eternity, as my mother wanted me to do? Or do I realize that all we want is care, and many of us are lacking that very thing in such crucial and foundational ways? And, if so, isn't it our responsibility to find empathy in our hearts for each other?

I have seen so many woman exercise love and unconditionality for a man in their life, while there would be a very clear and short boundary for real compassion for other women. We have the capacity, but maybe we feel, as women, femmes, or even folks who were assigned female at birth, that we don't deserve the compassion we crave because of the gendered construction of *who* is allotted that very compassion and care. Like society, we demean women every day in how we engage with each other. Our standards are low for men but high for women, and despite feminism, these are deeply embedded constructs we haven't yet healed from.

◎◎◎

During an ayahuasca retreat, I learned that in many communities' marijuana is representative of the divine feminine, and therefore her liberation is closely tied to the liberation of all women. I have also heard the analogous example of Mother Earth, of how her subjugation would only really end when we finally heal our wounds with women. Even symbolically, recreationally, we are misunderstood, misused, and misappropriated. This comes back to our interrelations as a species on this Earth. We have to see how we are communicating with her. Social and ecological accountability is what will create real shifts in the evolution of our species. Gender relationships are an important facet of understanding r(evolution). This is why turning to Indigenous organizations such as the Sogorea Te' Land Trust* and giving land back is an important part of our collective liberation. We must act now, holistically, on all fronts, to begin to mend the divide by choosing to change ourselves.

It's a huge mission to challenge how we extract and limit women and how that only serves a world and ecosystem built by and large by cis men, but it's a worthy endeavor to begin with yourself. Men have

* "Sogorea Te' Land Trust is an urban Indigenous women-led land trust based in the San Francisco Bay Area that facilitates the return of Indigenous land to Indigenous people." "Sogorea Te' Land Trust," accessed August 4, 2020, https://sogoreate-landtrust.org.

lived within a standard of *boys will be boys*, and we've accepted the pitfall of men, allowing them unconditional leeway toward sociopathy. Pretending as if power (in the traditional, masculine sense) is the only worthy and remarkable thing, when in actuality it's the most obvious and corrupting thing. We have thousands of years of history to prove how fickle we are, that despite mythologies and civilizations built fully showcasing our downfall, we repeat the propagation of power as if we have no free will, no recourse for evaluation nor change.

In the space of writing this book, I have been sad to see how many of my friends have been faced with darkness. To look toward the future now, as a thirty-two-year-old, is to see how easily we could have changed things if only we were willing to face ourselves. All this greed, for what? So corporations like BP can continue to spill oil in our oceans while we blame China for greenhouse gas emissions? So we can poison the rivers in Bangladesh black in order to have low-cut denim in every style? So more than one billion animals in Australia can die and forty-six million acres of land can burn just so we can live the lives we think we are owed?[6] As the renowned primatologist and anthropologist Jane Goodall once wrote, "Here we are, the most clever species ever to have lived. So how is it we can destroy the only planet we have?"

In late 2019, "Warning of a Climate Emergency" brought 11,258 scientists from 153 countries into agreement, calling for drastic changes to the economy, including a shift away "from GDP growth and the pursuit of affluence toward sustaining ecosystems and improving human well-being."[7] But this is a warning flag that scientists and environmentalists—not to mention land and water defenders—have *been* saying. In 2017, the Alliance of World Scientists made a similar call in a "Warning to Humanity" with 15,364 signatures calling for a complete change in how we exist. We have to change who we are and how we exist in this world, with each other and this Earth, to survive. But it takes a certain kind of cultural humility, Yunkaporta explains, which is "a useful exercise in understanding your role as an agent of sustainability in a complex system. It is difficult to relinquish the illusions of power and delusions of exceptionalism that come with privilege. But it is strangely liberating to realize your true status as a single node in a cooperative network. There is honor to be found in this role, and certain dignified agency."[8]

Philosopher Bayo Akomolafe explains that systems are a linguistic convenience, a construction, and how humans, try as we might, "tend to repeat and inscribe these crisis events even with our best efforts to resolve them because we're still stuck within the same epistemological space."[9] We're stuck in the construction of our tiny humanist ideas of what it means to be human. But that, by very definition, is an idea contorted by the white gaze, by the dominating, occupying culture's gaze. We are in the same loop for a bid for power, and within this state of spiritual crisis we remain until we accept our agency to release control of a system that has failed us and is still failing us. "The political spectrum itself is an illusion, suggesting that the only possible forms of social organization are liberalism, fascism, or socialism," Yunkaporta writes. "This limiting range of governance paradigms denies the existence of a myriad of forms of human society developed over eons of existence." Everything from manufactured adolescence (to domesticate people) to capitalism is a human construct, yet we criticize the existence of God and demonize spirit and magic because we fail to see them. We pledge allegiances to systems that are also invisible and yet prove only to divide and fail us, while adamantly convincing us otherwise. In tarot, the Tower card is one of my favorite cards. When you get it in a reading, it usually means whatever you're asking about is in a state of collapse. The only way out is through.

We are in a state of regeneration, but things are also collapsing. That much has to be accepted. What is "sustainable" in unsustainable living conditions? Who is wellness for in a world where wellness is not for all? How can we reconcile our embarrassing excessiveness? At the beginning of the pandemic, Arundhati Roy spoke of the pandemic as a portal; in the words of Akomolafe's *babalawo*, the wound is a portal. I have known this for a long time, that my wounds were my portals to a healed life—for what else is a healed life but utopia? These days I wonder more and more why we can't work toward utopia. To me, abolition is that, it's a belief that humans are messy, so we must create space for everyone to thrive. Enforcing barbaric systems is antithetical to a civil society, and to a just society. It's paradoxical to have a justice system that's lenient to some but not to others. These are all matters we should address as we move toward a collective future. "Why do we need police? Why do we need borders?" asks *The Red Deal*.[10] We can create something that supports all of our needs, a society where we can collectively reimagine who we want to be. "We need to go beyond critique and maybe edge towards experimental liminal spaces of transformation," Akomolafe tells us. We need to learn how to work together. "Every view point is useful, and it takes a wide diversity of views for any group to navigate this universe, let alone to act as custodians for it," Yunkaporta says.

◎ ◎ ◎

Many things have been clarified for me during the journey of writing this book. Thinking critically about who is well and who gets access to be well unfurled a question within myself, and I was asked to examine who I am more carefully. As a survivor, and as a highly traumatized person, I've found that saying what I feel, what my body feels, and what I need, has radically changed my relationship to myself and thus the world. To admit I needed more, that I wanted more care—care that was consistent and unconditional—felt like an impossible demand, yet I still wanted it deep in my heart. I believe it's possible to receive the care that I need, because it's the care I believe I give. There is no wellness for all, but there can be a society that accepts all people deserve to feel well and have access to whatever it is that will help them to get there. But we also have to understand that people have different needs and standards of wellness, and some do need more care than others because of what has happened to them. We must make room for people who have not traditionally taken up space to do so.

Disability rights activist Mia Mingus tells us that because we are building a reality that we have never seen before, it's imperative we flex in our vision of a truly inclusive world. But this also has to happen on many different fronts, and all of it has to be integrated for the future. Call it utopia, call it abolition, but there are futures awaiting to be made with our vision. So, what do we want? What do we want our world to become? If we could have anything, shouldn't we want more than this? Handing land back and allowing our Indigenous elders to become the true custodians of the land again, to guide us to the future, seems like a significant step toward that future. In that scenario, we protect this Earth and we save her from complete ecological collapse. Doesn't our species deserve a better ending than the alternative?

The Red Deal reminds us, "While making up only 5 percent of the world's population, Indigenous peoples protect 80 percent of the planet's biodiversity."[11] Juxtapose this against how white settlers own 96 percent of all agricultural lands and 98 percent of privately owned land in the United States. But imagine, as *The Red Deal* points out, if everyone's material needs were met, and their somatic and emotional needs were met, too. Imagine the world we could have. If there was no starvation, no homelessness, no alienation, where everyone was a relative and everyone had relatives. If we had trans health, nutrition, disability and elder care, reproductive health, dental health, behavioral and mental health, abortion, addiction, and HIV/AIDS treatment. "We can address these by organizing direct health services for unsheltered and poor relatives, including providing regular, free, and nutritious community meals; distributing Narcan to prevent opioid overdoses; offering needle exchanges; handing out free condoms. We can also organize campaigns to help new mothers. Contemporary reproductive health practices completely ignore follow-up care for new mothers (especially Black, Indigenous, brown, and poor mothers) are often marginalized in ableist activist cultures."[12] So that means we have to completely revolutionize how we organize care and center families, children, elders, and mothers.

I've been thinking of Australia a lot, because I have such a deep connection to *that* land. My home, my island home. In 2020, right before the pandemic, bushfires ravaged through Victoria, thousands of kilometers south of where I was raised. Three billion animals were killed or displaced and millions of acres were destroyed. The grief stayed in my body for months. My entire family is still there, while I am here. The water between us connects us, as does this Earth. My mother taught me to compost, she taught me how to care for the land. These days I wonder what a miracle that is, and how complicated love can be. It's powerful to see her clearly, still blurry from the pain, but less wounded. Forgiving her is an act of composting grief into this land, into the Earth. I was taught to do that in ceremony. When I asked how this Earth could hold more pain, I was reminded that she is our mother. And like the gracious mother that she is, she can hold immense amounts of pain. Her ferocity comes from wisdom.

These are impossible times, but that's why they deserve impossible dreaming and visioning. In an interview, American whistleblower Edward Snowden explained: "When we think about the civil rights movement, when we think about every social progress that's happened throughout history, going all the way back to the Renaissance, going back to people thinking about heretical ideas—'Hey, maybe the world is not flat'—even making these arguments, challenging conventions, challenging the structures of law on any given day itself is a violation of law."[13] But we can and will have to change. Healing our relationship to femininity, to women, to this Earth, will shift our story as a species immensely. A few years ago, I started contemplating: Tending to ourselves, healing ourselves, has an energetic impact on the people around us. Understanding this means healing our relationships between women and femmes—through accessing and prioritizing care—as well as working through those societal standards that rely on disconnecting women from each other. The more we turn on each other, upholding men, the more they will continue to dominate. But what has all this domination led to? Why can't we live together, in peace?

Witness how you treat other women. Do you put them on impossible pedestals? How do you react when another woman fails you? Hurts you? Is messy? Do you have to make another woman feel small in order to feel big? How do you treat a woman you don't like? Is it the same as how you treat a man you don't like? Be honest. I've had to acknowledge my own deep-seated hypocrisies, foibles, and insecurities, and how that has sometimes encouraged jealousy or resentment. I've had to have that honest communication with myself, so now when that happens, I let the feeling of jealousy guide me to the real heart of the matter. Every facet of being in right relationship means engaging truthfully with oneself. Restructuring how we relate to the feminine will change us, empower us. We deserve to have each other's love and care. We deserve that kind of holistic liberation.

CONCLUSION: ON SACRED RECIPROCITY

When we love the Earth we are able to love
ourselves more fully.

—bell hooks

Recently I came across the phrase *meden agan,* which was the second maxim inscribed on the forecourt of the Temple of Apollo at Delphi. It means "nothing in excess." My father used to tell me, and still does when I let him, "Baba, everything in moderation." If he had a phrase, this would be it.

As a child, when he'd tell me this, I'd only think of restrictions. I thought that a good life was opposite to the one that I had—a sheltered, fractured, hostile life—a life of lack. So, a good life meant plenty, it meant indulgence. In the last few months, I've considered this in the realms of my own body. Learning to stay still with myself, cultivating that ear, has meant that I've had to listen when my body has shut down against alcohol, dairy, and sugar. Things I feel I am programmed to give it, through commercials and capitalism, because a good life, to me, meant excess. When I want to indulge, rarely do I think of something that will make me feel good, but rather I'll crave the prescribed idea of that thing. This is why American advertising is so insidious—it relies on selling you the life it wants you to want.

For me, my want for things came from a place of lack. I felt as if I had nothing, and I desperately wanted something. Remember that feast scene in *A Little Princess?* I felt I was her always, trying to make something out of emptiness. But, though I was make-shifting a life of joy, of a family, a feast, a good life, it was one that Hollywood sold me. Not what was mine, not what was in my heart, not my own dreams.

Dreaming is extremely important for the revolution. Believing in possibility, what can come and what *should* come requires all of us to dream for the future, for more.

Brazilian architect and urbanist Paulo Tavares writes in the essay "The Political Nature of the Forest: A Botanic Archaeology of Genocide" that the Amazon was always seen as a "void space characterized by chronic lack"—that it only represented "demographic emptiness, technological underdevelopment, economic stagnation, territorial isolation,"[1] within the construction of the colonial imagination. There wasn't an understanding that a relationship to the land, to each other, had value. There was no humility in understanding that another way of life could have wisdom for the colonizer. It reminds me of what Vandana Shiva spoke to about the forests of India, of my grandparents' homeland. When they looked at the forests of Bengal, all the colonizers saw and believed was, again, a lack. They built an entire ethos on lack, and hundreds of years of European philosophy shows the disregard of human life, and how that metabolized in the consciousness of whoever was struck by colonial thought. So, basically, everyone. White men want to barrage everyone with their existentialism, but maybe the bleakness of their soul lies in their incapacity to appreciate this Earth, this bounty. They don't see how she deserves their protection. God is here, we just have to open our hearts to each other.

Indigenous peoples of the world have lived sustainably in accordance with the land, because among a plethora of land wisdom, they have always understood the nature of sacred reciprocity. When an ayahuasca teacher first introduced me to this concept, I felt a ripple of bright recognition run through me like a gentle wave. Sacred reciprocity is a concept that relies on an understanding that there is something inherently spiritual in your exchange with others, with the planet, with the trees, the birds, the spectacular wildlife of this lush and verdant planet. Do you ever get high and watch nature documentaries? You know how you can feel the rupture of the planet's pull next to the grandiosity of a peacock's mating dance? Or the hum of the ocean's wind as a dolphin pirouettes from water into the sky? This Earth, this great big Earth, with her mountainous icy peaks, her limestone edges and creeks that glimmer with pearly moss, every part of this planet has a secret waiting to unravel. Instead of spending billions of dollars on endless military, mindlessly expanding an empire for no cause—what if that money was invested back into the land, into protecting and defending her?

"We must reclaim Indigenous intellectual traditions of the Western Hemisphere, which have some of the most advanced technology in human history."[2] We have to understand that the Earth has logic, and magic, that we don't understand, but through cultivating and reinvesting in her, by cultivating and reinvesting in ourselves, we create new paradigms of being. We have to understand the symbiosis between species, that there is deep wisdom in how a butterfly breaks through its cocoon and how a bee collects pollen. These systems are interdependent, so we must move with them, with the land. "Maintain diverse languages, cultures, and systems that reflect the ecosystems of the shifting landscapes you inhabit over time," Yunkaporta guides us with his yarn, his deep Indigenous knowledge and research. "There is an undeniable pattern in the sum total of all these old stories from around the world, indicating that sedentary lifestyles and cultures that do not move with the land or mimic land based networks in their social systems do not transition well through apocalyptic moments."[3] The more we work against the land, the harder it will be for us. It's important to understand that ecological dependency is paramount. According to Tavares, ecocide and genocide are closely linked, often a mirror of one another, in the fight for land sovereignty. Understanding other ecologies we share on this Earth, understanding that it's important that we start thinking with these other species in mind, are the steps that need to be taken by all of us. Learning we are connected to the Earth, and everything she embodies, means we take better care of it all, understanding that we have a responsibility for not only our own survival, but hers, and theirs. It's a mutual responsibility to plants, to animals, to the water, vast and sublime. Sacred reciprocity is an ethos we must embody toward everything we share this planet with. To echo Jane Goodall, humans are the only ruinous species on the planet. Is this all our intelligence has amounted to?

Scarcity is linked to capitalism because we need to feel scarce in order to want more. It's a deflection device, created to convince us of our immortality, but when we die, we will be put back into this Earth. Isn't it beautiful that we can honor what will hold us in the afterlife? There's a sacred reciprocity in that, as well. So, what do we really want in this lifetime, for ourselves, truly? Kimmerer believes what we really want is to be valued, "earning the regard of your neighbors for the quality of your character, not the quantity of your possessions; what you give, not what you have."[4] Looking at my own life, my own deep cravings, I would agree. It's about reorienting the family unit and understanding, again, that we are all we have. The connections in this lifetime are what mature us, teach us, sustain us, and when we are in sacred reciprocity with each other, honoring each other's pasts, experiences, and needs, we can better help facilitate true healing for humanity and the planet.

I know my relationship to my mother is difficult, yet it is also a portal. It's a wound that's become bigger than me, and it's given me a mission—to make art to heal. Forgiveness is an important component here. For my body, and for hers. Understanding that this was my path, and that was hers, has helped lift the load that trauma weighs. The exceptionalism in devastating traumatic events creates a narrative of victimization that I needed to pull myself out of. All these years through this journey, I've asked: What's standing in my way of forgiveness? On a societal level, it's important to ask the same question: What is standing in the way of our liberation? To love ourselves and others? Sacred reciprocity is tied to accountability of yourself, because you have a moral code to yourself. When you take only as much as you give, it gets easier to see those who do the same. If we could trust one another, then maybe we'd consider setting a higher standard for ourselves. It's no surprise that a society that refuses to take accountability perpetuates a people who do the same.

"The system that we have is so based in, 'You're a bad person, so bad things deserve to happen to you,'" Mia Mingus tells us. "Rather than, 'We are all human, and human beings make mistakes, and we are all flawed, and we are all living in incredibly violent conditions.'"[5] Abolition and transformative justice are acutely linked to the land, but that imprint needs to happen among us humans more. We need to take ourselves as custodians of this Earth seriously and understand how grave these circumstances are. Evolutionary biologist David Sloan Wilson has found that competition makes sense only when it's about the individual; however, when it comes to a group, cooperation is a better model, not only for surviving but for thriving. "There is symbiosis at every single level of living things," Richard Powers writes, "and you cannot compete in a zero-sum game with creatures upon whom your existence depends."[6]

For the skeptical, Kimmerer responds: "You might rightly observe that we no longer live in small, insular societies, where generosity and mutual esteem structure our relations. But we could." It is within our power to develop an interconnectedness outside the market economy. "Intentional communities of mutual self-reliance and reciprocity are the wave of the future, and their currency is sharing."[7] There is vast potential here, and that's what we should remember. There is excitement here, newness—but above all, there's accessibility, for *all*. Not accessibility for a select few at the expense of most others. Creating local food economies by investing in community gardens or local farms (especially Indigenous- or Black-owned ones) instead of expending energy and resources on trade is one way of the future. Creating stronger mutual aid networks that protect everyone, divesting from capitalism by degrowing and redistributing wealth are ways toward our collective future.

For cases where imported goods can't be substituted, we can minimize trade with the West and opt instead to trade with the Global South. This is what Egyptian economist Samir Amin calls "de-linking" from unequal exchange. In 1978, Japanese farmer Masanobu Fukuoka advocated for a "do-nothing" type farming, calling for curiosity, openness, and a willingness to fail, so that we can learn to trust ourselves, but more the cycles of the Earth. His farming practices created no pollution and didn't use fossil fuels. There are ways we haven't yet tried, ways to be in accordance with every facet of the Earth. This means taking responsibility for your usage, but also your actions. Intentionality, now, needs to be sourced from deep within.

I want to live in accordance with sacred reciprocity, to exist there, in that heart-centeredness. It's wild that people, society, want to convince you out of the power of your heart. Of the will it takes to listen, but the hunger that is satiated when you move with integrity, that there's nothing quite like it. By using sacred medicines like marijuana, mushrooms, and Grandmother ayahuasca, I have found deep healing and understand that plants are gateways to the divine. I honor these medicines and the traditional custodians that brought us this knowledge, the wisdom keepers, who did nothing for profit but only for the love of the Earth.

◎ ◎ ◎

Something changed in me when I began to let myself absorb the spirit of this great planet. It happened soon after I sat with Grandmother ayahuasca for the first time. Completely unbridled, she cracked me open, reminding me that to be porous, to be vulnerable, is the most generous thing you can offer. It's fascinating how having gratitude for this life, whatever this life has brought you, brings profundity to what surrounds us. Ojibwe elder and activist Winona LaDuke says that Indigenous practice is "about recovering that which the Creator gave us as instructions, and then walking that path." To me, there's nothing more honorable than that. To find a devotion so deep toward God that gratitude is all that remains. We are here, we are here, together, and that matters.

I don't need to convince anyone of anything, and yet this beautiful tragic life I've been given has brought such awakening. My entire life I have resented and resisted what stood before me and behind me, like a wide, nauseating ache. I have felt so at the behest of my life, but that changed when I started praying. When I committed to prayer every day—not on a *janamaz* as I was taught, but something more informal, between God and me—something like a spell lifted. I never had to abandon God, and we never had to abandon this planet.

I have lived many years in pain, but I have a new drive to move through these machinations of being. There is sacredness here, there always was, it's about seeing it with fresh eyes and remembering everything can be a teaching. "It is the same for the woman in exile. If she is an ugly duckling, if she is unmothered, her instincts have not been sharpened. She learns instead by trial and error. Usually many trials; many, many errors. But there is hope, for you see, the exile never gives up. She keeps going till she finds the guide, the scent, till she finds the trail, till she finds home,"[8] Estés writes with necessity in *Women Who Run with the Wolves*.

Finally, I am finding home in myself, by being honest, true. Wanting to align with integrity in myself meant accepting the totality of my being—the hurt and the joy, the sacredness and the profane. I was going to choose to be bold, in myself, in my own intuitive understanding of self. I see how revolutionary this act is. To love yourself when others have loved you so poorly is a reminder that anything can be shifted into something meaningful. Sometimes it's about interpretation, remembering sacred reciprocity, and moving within that framework. Understanding it's not my job to serve anybody has shown me care could mean mutuality. By determining my own needs, I am finding home on this Earth by declaring myself worthy of a good life, but not at the expense of another's. I hope to protect every human on this planet, even the ones I dislike, because I cherish human life. I cherish our time here, together, on this mortal plane, and with that reminder I look toward something bigger than my own needs. It's learning how to contain multitudes; to hold many things at once is a generous act that we owe to each other. Forgiveness and generosity—more of us have to move with that understanding because there's something vast at stake here, and thus worth fighting for. When I found worthiness in myself, in protecting myself, it got easier to think about protecting the Earth. Wellness for all is a worthy pursuit. It means true planetary healing. We have to believe it's possible.

These days, I look to the trees, my ancestors. I feel their pulse in my skin, in my heart, in the beat of my breath as I walk past the reservoir near my house. When I was a teenager, my mother's house also had a reservoir just behind the bush, and I would go there when I was hurt, to nurse my wounds with water and moss. This Earth has always given me recovery, my tears connected me to her rivers, to the creeks that gushed with pebbles, oceans that generated such stillness and ferocity. I found such a holy sanctuary there, in the pockets of nature's grasp.

I always assumed I didn't have a mother, but now I look back and I realize my mother was this Earth, and in moments of tenderness, loneliness, and even awe, she was always there to hold me, to tend to me, to brace me with her grounding. I owe everything to this Earth for my survival. The trees are my kin, they listen to me weep, they hear my gratitude as I pray to their roots for patience and grace, all the wisdom that they have taught me. I find great mercy with them, with the eucalyptus and the lilac and the Meyer lemon trees that surround me. They root me through my unrootedness so much that all I can say is thank you. My gratitude for this Earth is endless. What a privilege it is to be with her in this lifetime and to have her guidance and care. It's so humbling to be alive. And I will never, ever stop saying thank you. For my life, for everything. Thank you.

ACKNOWLEDGMENTS

First, thank you to the people, spirits, elders, guides, and ancestors who held it down for me.

For the people who showed me I'm easy to love. Thank you.

Thank you to the Harper Wave team—especially to Julie Will and Emma Kupor. My gratitude to Arsh Raziuddin for the beautiful cover. Thank you also to Monika Woods!

Thank you to my darling friends, some of whom were also early readers: Zeba Blay, Prinita Thevarajah, Fatimah Asghar, Safia Elhillo, James J Robinson, Mimi Zhu, Angela Dimayuga, Ganavya Doraiswamy, Emily Margo, Zenat Begum, Konrad Konczak-Islam, David Ehrlich, Raven Leilani, Clémence Polès, Alês Kot, Fariba Alam, Megan Williams, Marlee Grace, Raneen Bukhari, Hisham Fageeh, Meital Yaniv, Arabelle Sicardi, Shriya Samavai, Hassan Rahim, and Alex Farbstein.

Thank you to my students and everyone who has supported my teaching. I love you all; you made me realize my work has value. Thank you for being such a beautiful reflection of my heart, and for proving that the possibilities of manifesting community are immense. Thank you to everyone who's ever read and supported me. I feel you in my bones. Thank you to DegrowthNYC for showing me that there are others out there like me who want to fight for revolution . . . and who are actively reimagining ways of being together.

I have immense respect for the writers who taught me how to find my voice. I want to give my thanks to the elders who died while

I wrote this manuscript: to Etel Adnan, bell hooks, Nawal El Saadawi, Thich Nhat Hanh, and Joan Didion. It's been wild to lose so many teachers who paved the way, but I am grateful to have read your words, to be guided toward the truth with the torches that you bore. Thank you immensely to Sebene Selassie, Robin Wall Kimmerer, Arundhati Roy, Vandana Shiva, Gabor Maté, Judith Herman, Leah Lakshmi Piepzna-Samarasinha, and Amita Swadhin, in particular, for being beacons of light through this dark path toward liberation. You were all pilgrimages as I wrote this book. Thank you also to James Baldwin, Audre Lorde, Susan Sontag, and June Jordan for being the blueprint.

Thank you to the Muslims and South Asians who support me. I write for us.

Thank you to my family. To my lineage, to my ancestors. To my grandparents on both my sides, to Abdul, my twin. I hear your call, I'm listening. Thank you to my mother for bringing me here to do this kind of service. I am grateful for the life you gave me and I am humbled by you, even if I don't feel it all the time. We are bound together in this mysterious life. Thank you to my father for seeing the magic in me and nurturing it as best you knew how. Thank you for radicalizing me young. Thank you for making me question authority. Thank you for making me challenge white supremacy, for never giving in, for being an example of a man. Of a father. I love you so much and I am so grateful to be your daughter. To Samia, my sister. There are so many parts of me that are just parts I took from you. Thank you for guiding me, for being generous with your love and for showing me that it's okay to push back, to fight for yourself. We are on such a specific journey together. I am proud of us for surviving so much.

Last, my gratitude to the land. To the sacredness of the work of all land and water defenders. I am beside you and behind you. Thank you for the sacred medicines. For the wisdom keepers who have shared their knowledge via the Earth's immense wisdom. Thank you to Grandmother ayahuasca, in particular, for your call, for your awakening, for your healing, for your love. I'm so humbled by your teachings. You changed my life.

Thank you, God, for everything. Thank you.

Onward toward revolution,

Fariha, spring 2022

NOTES

Epigraph

1. Joanna Macy, "Entering the Bardo," *Emergence,* October 5, 2021, https://emergencemagazine.org/op_ed/entering-the-bardo/.

Introduction

1. Zoë Carpenter, "What's Killing America's Black Infants?" *The Nation,* December 23, 2019, https://www.thenation.com/article/archive/whats-killing-americas-black-infants/.
2. Resmaa Menakem, *My Grandmother's Hands: Racialized Trauma and the Pathway to Mending Our Hearts and Bodies* (Penguin Books, 2021), 9.
3. June Jordan, *Civil Wars* (Simon & Schuster, 1995), 12.
4. Anne Boyer, *The Undying: Pain, Vulnerability, Mortality, Medicine, Art, Time, Dreams, Data, Exhaustion, Cancer, and Care* (Picador, 2020), 79.
5. Robin Wall Kimmerer, *Braiding Sweetgrass* (Tantor Media, Inc., 2016), 47.
6. Kimmerer, *Braiding Sweetgrass,* 338.

Chapter 1: On the Mind

1. Kimberlé Crenshaw, host. "Storytelling While Black and Female: Conjuring Beautiful Experiments." *Under the Blacklight,* August 6, 2020, https://soundcloud.com/intersectionality-matters/24-storytelling-while-black-and-female-conjuring-beautiful-experiments-in-past-and-future-worlds.
2. Gabor Maté, *When the Body Says No: Exploring the Stress-Disease Connection* (Vermilion, 2019), 16.
3. David Wojnarowicz, *Close to the Knives: A Memoir of Disintegration* (Canongate Books, 2017).
4. Maté, "The Bermuda Triangle," *When the Body Says No,* 9.
5. Alaka Wali, "Democracy and the Iroquois Constitution," November 1, 2016, https://www.fieldmuseum.org/blog/democracy-and-iroquois-constitution (accessed 2/12).
6. "Storytelling and Survivorship with Amita Swadhin," Studio Ānanda, https://www.studioananda.space/2021/09/01/storytelling-and-survivorship-with-amita-swadhin/.
7. Maté, "The Bermuda Triangle," *When the Body Says No,* 78.
8. Maté, "The Little Girl Too Good to Be True," *When the Body Says No,* 19.

9. Elaine Scarry, *The Body in Pain: The Making and Unmaking of the World* (Oxford University Press, 1985).
10. Anne Boyer, "What Is the Language of Pain? An Ugly Gathering of Adjectives," *Literary Hub*, September 16, 2019. https://lithub.com/anne-boyer-what-is-the-language-of-pain/.
11. Susan M. Johnson, *Hold Me Tight: Seven Conversations for a Lifetime of Love* (Little, Brown, 2010), 34.

Chapter 2: On Meditation

1. Matthew Thorpe and Rachael Link, "12 Benefits of Meditation," *Healthline*, October 27, 2020. https://www.healthline.com/nutrition/12-benefits-of-meditation.
2. Chris Brennan, host, "Queen Buran, Astrologer in 9th Century Baghdad," *The Astrology Podcast*, March 12, 2021. https://theastrologypodcast.com/2021/03/12/queen-buran-astrologer-in-9th-century-baghdad/.
3. "Finding the Cure for Pain in the Pain with Shadi Sankary," Studio Ānanda, September 1, 2021. https://www.studioananda.space/2021/09/01/finding-the-cure-for-pain-in-the-pain-with-shadi-sankary/.
4. Budd L. Hall and Rajesh Tandon, "Decolonization of Knowledge, Epistemicide, Participatory Research and Higher Education," *Research for All* 1, no. 1 (2017): 6–19. https://doi.org/10.18546/RFA.01.1.02.
5. Hall and Tandon, "Decolonization of Knowledge, Epistemicide, Participatory Research and Higher Education," 6–19.
6. Robert Puff, "An Overview of Meditation: Its Origins and Traditions," *Psychology Today*, July 7, 2013. https://www.psychologytoday.com/intl/blog/meditation-modern-life/201307/overview-meditation-its-origins-and-traditions.
7. R. K. Kaushik, *Architect of Human Destiny?: Who Brings about Peace or Chaos?* (Kalpaz Publications, 2003), 6.
8. David Dale Holmes, "A Buddhist View on Caste and Equality," Buddhistdoor Global, June 15, 2021. https://www.buddhistdoor.net/features/a-buddhist-view-on-caste-and-equality/.
9. Shridhar Sharma, "Psychiatry, Colonialism and Indian Civilization: A Historical Appraisal," *Indian Journal of Psychiatry 48, no. 2 (2006)*: 109–12. https://doi.org/10.4103/0019–5545.31600.
10. Omer Aziz, "Blighted by Empire: What the British Did to India," *Los Angeles Review of Books*, September 1, 2018. https://lareviewofbooks.org/article/blighted-by-empire-what-the-british-did-to-india/.
11. Dadabhai Naoroji, "Poverty and Un-British Rule in India," Internet Archive, Swan Sonnenschein, 1901. https://archive.org/details/povertyunbritish00naoruoft/page/n3/mode/2up.
12. Dadabhai Naoroji, *Poverty of India* (Forgotten Books, 2018).
13. Sharma, "Psychiatry, Colonialism and Indian Civilization," 109–12.
14. Sharma, "Psychiatry, Colonialism and Indian Civilization," 109–12.

15. David Forbes, "How Capitalism Captured the Mindfulness Industry," *Guardian*, April 16, 2019. https://www.theguardian.com/lifeandstyle/2019/apr/16/how-capitalism-captured-the-mindfulness-industry.
16. Forbes, "How Capitalism Captured the Mindfulness Industry."
17. Ron Purser and David Loy, "Beyond McMindfulness," *HuffPost*, August 31, 2013. https://www.huffpost.com/entry/beyond-mcmindfulness_b_3519289.
18. Sharma, "Psychiatry, Colonialism and Indian Civilization,"109–12.
19. Ronald Purser, "From Neoliberal to Social Mindfulness," Powell's Books, July 12, 2019. https://www.powells.com/post/original-essays/from-neoliberal-to-social-mindfulness.
20. Ashley Ross, "How Meditation Went Mainstream," *Time*, March 9, 2016. https://time.com/4246928/meditation-history-buddhism/.
21. Johanna Slater and Niha Masi, "India Passes Controversial Citizenship Law Excluding Muslim Migrants, *Washington Post*, December 11, 2019.
22. B. R. Ambedkar, *Annihilation of Caste* (New Delhi: Rupa, 2018).
23. Arundhati Roy, *The End of Imagination* (Haymarket Books, 2016), 5.
24. Trinh T. Minh-Ha, *When the Moon Waxes Red: Representation, Gender and Cultural Politics* (Routledge, 2013), 20.
25. Rabindranath Tagore, *Nationalism* (Penguin Books, 2017).
26. Sir Rabindranath Tagore, "Works of Tagore from the Modern Review, 1909–24/The Spirit of Japan," Wikisource. https://en.wikisource.org/wiki/Works_of_Tagore_from_the_Modern_Review,_1909–24/The_Spirit_of_Japan.
27. Purser and Loy, "Beyond McMindfulness."
28. David Kortava, "Lost in Thought," *Harper's Magazine*, April 2021. https://harpers.org/archive/2021/04/lost-in-thought-psychological-risks-of-meditation/.
29. Michael Yellow Bird, *Neurodecolonization and Indigenous Mindfulness*. https://www.indigenousmindfulness.com/about.
30. Michael Yellow Bird, "Decolonizing the Mind: Using Mindfulness Research and Traditional Indigenous Ceremonies to Delete the Neural Networks of Colonialism." Accessed February 15, 2022. http://www.aihec.org/our-stories/docs/BehavioralHealth/2016/NeurodecolonizationMindfulness_YellowBird.pdf.
31. Tagore, "Works of Tagore from the Modern Review, 1909–24/The Spirit of Japan."

Chapter 3: On Intuition and Unseen Things

1. Masaru Emoto, *The Hidden Messages in Water* (Beyond Words, 2004).
2. Dwight Garner, "Inside the List," *New York Times*, March 13, 2005, https://www.nytimes.com/2005/03/13/books/review/inside-the-list.html.
3. "Clean Living Movement," Wikipedia, accessed September 25, 2021, https://en.wikipedia.org/wiki/Clean_living_movement.

4. Sabrina Strings, *Fearing the Black Body: The Racial Origins of Fat Phobia* (New York University Press, 2019), 196.
5. Lisa Ellis, "Snapshot of the American Pharmaceutical Industry," Harvard T.H. Chan School of Public Health, October 28, 2019. https://www.hsph.harvard.edu/ecpe/snapshot-of-the-american-pharmaceutical-industry/.
6. Cheryl Wischhover, "Can Wellness Be Scientific?" *The Cut*, July 19, 2016. https://www.thecut.com/swellness/2016/07/can-wellness-be-scientific.html.
7. Sebene Selassie, *You Belong: A Call for Connection* (HarperOne, 2021), 21.
8. Clarissa Pinkola Estés, *Women Who Run with the Wolves: Myths and Stories of the Wild Woman Archetype* (Ballantine Books, 1995), 8, 47.
9. Estés, Women *Who Run with the Wolves*, 8.
10. Mariana Alessandri, "The Gender Politics of Fasting," *New York Times*, January 14, 2019. https://www.nytimes.com/2019/01/14/opinion/fasting-gender-politics.html.
11. Alessandri, "The Gender Politics of Fasting."
12. Alessandri, "The Gender Politics of Fasting."
13. Cheryl Wischhover, "Can Wellness Be Scientific?"
14. Alessandri, "The Gender Politics of Fasting."
15. Garner, "Inside the List."
16. William Reville, "The Pseudoscience of Creating Beautiful (or Ugly) Water," *Irish Times*, February 17, 2013, https://www.irishtimes.com/news/science/the-pseudoscience-of-creating-beautiful-or-ugly-water-1.574583.
17. Franklin Edgerton, *Buddhist Hybrid Sanskrit Grammar and Dictionary* (Yale University Press, 1972), 221.
18. Mimi Thi Nguyen, *The Gift of Freedom: War, Debt, and Other Refugee Passages* (Duke University Press, 2012), 11.
19. Selassie, *You Belong*, 47.
20. Ivan Illich, *Limits to Medicine: Medical Nemesis, the Expropriation of Health* (M. Boyars, 2013).
21. Gabor Maté, "Opinion: Time to Stress Cancer Rates Among the Young," *The Globe and Mail*, April 24, 2002. https://www.theglobeandmail.com/opinion/time-to-stress-cancer-rates-among-the-young/article754321/.
22. Anne Boyer, *The Undying: Pain, Vulnerability, Mortality, Medicine, Art, Time, Dreams, Data, Exhaustion, Cancer, and Care* (Farrar, Straus & Giroux, 2019), 123.
23. Gabor Maté, "The Bermuda Triangle," *When the Body Says No: Exploring the Stress-Disease Connection* (Vermilion, 2019), 9.
24. "Missing and Murdered Indigenous Women and Girls," *National Indigenous Women's Resource Center*. https://www.niwrc.org/mmiwg-awareness.
25. Maté, "The Bermuda Triangle," *When the Body Says No*, 9.

26. Selassie, *You Belong*, 48.
27. Penelope K. Trickett, Jennie G. Noll, and Frank W. Putnam, "The Impact of Sexual Abuse on Female Development: Lessons from a Multigenerational, Longitudinal Research Study," *Development and Psychopathology* 23, no. 2 (2011): 453–76. https://doi.org/10.1017/S0954579411000174.
28. *Diagnostic and Statistical Manual of Mental Disorders: DSM-5*. Arlington, VA: American Psychiatric Association, 2017.
29. Silvia Federici, *Caliban and the Witch: Women, the Body and Primitive Accumulation* (Autonomedia, 2014), 100.
30. Brennan, Chris, host, "Queen Buran, Astrologer in 9th Century Baghdad." *The Astrology Podcast*, March 12, 2021. https://theastrologypodcast.com/2021/03/12/queen-buran-astrologer-in-9th-century-baghdad/.
31. "Mind–Body Dualism," Wikipedia. https://en.wikipedia.org/wiki/Mind%E2%80%93body_dualism.
32. Leah Lakshmi Piepzna-Samarasinha, *Care Work: Dreaming Disability Justice* (Arsenal Pulp Press, 2021), 71.
33. Estés, *Women Who Run with the Wolves*, 312.
34. Piepzna-Samarasinha, *Care Work*, 66.

Chapter 4: On the Frustrating Pitfalls of Healing

1. Arundhati Roy, *The End of Imagination* (Haymarket Books, 2016), 263.
2. Roy, *The End of Imagination*, 265.
3. C. J. Hauser, "The Crane Wife," *Paris Review*, August 19, 2019. https://www.theparisreview.org/blog/2019/07/16/the-crane-wife/.
4. Leah Lakshmi Piepzna-Samarasinha, *Care Work: Dreaming Disability Justice* (Arsenal Pulp Press, 2021), 215.
5. Gabor Maté, *When the Body Says No: Exploring the Stress-Disease Connection* (Vermilion, 2019), 3.

Chapter 5: Introduction to the Body

1. Lama Rod Owens, *Love and Rage: The Path of Liberation through Anger* (North Atlantic Books, 2020), 19.
2. Jackie Wang, *Carceral Capitalism* (Semiotext(e), 2018), 14.
3. Kimberlé Crenshaw, host, "What's the Matter with Georgia? Virus, Voting, & Vigilantism the Peach State," *Under the Blacklight*, May 13, 2020. https://soundcloud.com/intersectionality-matters/17-under-the-blacklight-virus-voting-vigilantism-in-georgia.
4. Anand Giridharadas, *Winners Take All: The Elite Charade of Changing the World* (Alfred A. Knopf, 2018), 3.
5. David Wojnarowicz, *Close to the Knives: A Memoir of Disintegration* (Canongate Books), 2017.

6. Silvia Federici, *Caliban and the Witch: Women, the Body and Primitive Accumulation*. (Autonomedia, 2014), 15.
7. Federici, *Caliban and the Witch*, 14.
8. Robin Wall Kimmerer, *Braiding Sweetgrass* (Tantor Media, 2016), 304.
9. Kimmerer, *Braiding Sweetgrass*, 308.
10. Giridharadas, *Winners Take All*, 8.
11. Giridharadas, *Winners Take All*, 34.
12. Alice Slater, "The US Has Military Bases in 80 Countries. All of Them Must Close," *The Nation*, January 25, 2018. https://www.thenation.com/article/archive/the-us-has-military-bases-in-172-countries-all-of-them-must-close/.
13. Vivek Bald, "Unsettling the Nation from the Land: A Conversation with Manu Karuka," Asian American Writers' Workshop, *The Margins*, May 29, 2019. https://aaww.org/unsettling-the-nation-from-the-land-a-conversation-with-manu-karuka/.
14. Bald, "Unsettling the Nation from the Land."
15. Bald, "Unsettling the Nation from the Land."
16. Federici, *Caliban and the Witch*, 11.
17. Julietta Singh, *Unthinking Mastery: Dehumanism and Decolonial Entanglements* (Duke University Press, 2018), 34.
18. Paulo Freire, *Pedagogy of the Oppressed* (Continuum, 2005), 44.
19. Saidiya V. Hartman, *Scenes of Subjection: Terror, Slavery, and Self-Making in Nineteenth-Century America* (Oxford University Press, 2010), 11.
20. Hartman, *Scenes of Subjection*, 53.
21. Hartman, *Scenes of Subjection*, 52.
22. Hartman, *Scenes of Subjection*, 53.
23. François Bernier, "Voyage dans les États du Grand Mogol," introduction de France Bhattacharya (Arthème Fayard, 1981).
24. Siep Stuurman, "François Bernier and the Invention of Racial Classification," *History Workshop Journal* 50 (2000): 1–21. http://www.jstor.org/stable/4289688.
25. Anna-Sophie Springer and Paulo Tavares, "The Political Nature of the Forest: A Botanic Archaeology of Genocide." *The Word for World Is Still Forest* (K. Verlag, 2017).
26. Pierre H. Boulle, "Francois Bernier and the Origins of the Modern Concept of Race," *The Color of Liberty: Histories of Race in France*, Sue Peabody and Tyler Stovell, eds. (Duke University Press, 2003), 11–19.
27. Sabrina Strings, *Fearing the Black Body: The Racial Origins of Fat Phobia* (New York University Press, 2019), 68.
28. Vinson Cunningham, "The Future of L.A. Is Here. Robin D.G. Kelley's Radical Imagination Shows Us the Way," *Los Angeles Times*, March 17, 2021. https://www.latimes.com/lifestyle/image/story/2021-03-17/robin-dg-kelley-black-marxism-protests-la-politics.

Chapter 6: On Body Dysmorphia

1. Julietta Singh, *No Archive Will Restore You* (Punctum Books, 2018), 91.
2. Sebene Selassie, *You Belong: A Call for Connection* (HarperOne, 2021), 73.
3. Gabor Maté, "The Bermuda Triangle," *When the Body Says No: Exploring the Stress-Disease Connection* (Vermilion, 2019), 203.
4. Selassie, *You Belong*, 99.
5. Maté, *When the Body Says No*, 192.
6. Maté, "The Bermuda Triangle," *When the Body Says No*, 27.
7. Maté, *When the Body Says No*, 208.
8. Siep Stuurman, "François Bernier and the Invention of Racial Classification," *History Workshop Journal* 50 (2000): 1–21. http://www.jstor.org/stable/4289688.
9. "Enmeshment," Wikipedia, January 15, 2022. https://en.wikipedia.org/wiki/Enmeshment.
10. Adrienne Maree Brown, *Pleasure Activism: The Politics of Feeling Good* (AK Press, 2019), 289.
11. Julietta Singh, *Unthinking Mastery: Dehumanism and Decolonial Entanglements* (Duke University Press, 2018), 152.
12. Sabrina Strings, *Fearing the Black Body: The Racial Origins of Fat Phobia* (New York: New York University Press, 2019), 6.
13. Maté, *When the Body Says No*, 110.
14. Maté, *When the Body Says No*, 223.
15. Brown, *Pleasure Activism*, 305.
16. Audre Lorde, *The Audre Lorde Compendium* (Pandora, 1996), 101.
17. Brown, *Pleasure Activism*, 440.

Chapter 7: On White People Co-opting Yoga

1. Resmaa Menakem, *My Grandmother's Hands: Racialized Trauma and the Pathway to Mending Our Hearts and Bodies* (Penguin Books, 2021), 9.
2. Swami Vivekananda, *Karma Yoga: The Yoga of Action* (Vedanta Press—Advaita
3. Ashrama, 2007), 49.
4. Meera Navlakha, "For Indians in the Grip of a COVID-19 Crisis, Twitter Is a Desperate Last Resort," *gal-dem*, April 23, 2021. https://gal-dem.com/india-covid-19-twitter/.
5. Prinita Thevarajah, "How Casteism Manifests in Yoga and Why It's a Problem," *Byrdie*, September 9, 2021. https://www.byrdie.com/casteism-in-yoga-5119378.
6. Thevarajah, "How Casteism Manifests in Yoga and Why It's a Problem."
7. Thevarajah, "How Casteism Manifests in Yoga and Why It's a Problem."
8. Vivekananda, *Karma Yoga*, 54.
9. Davis Gordon White, "Yoga, Brief History of an Idea." Princeton University

Press. Accessed February 15, 2022. https://assets.press.princeton.edu/chapters/i9565.pdf.
10. White, "Yoga, Brief History of an Idea."
11. White, "Yoga, Brief History of an Idea."
12. David Gordon White, ed., *Yoga in Practice* (Princeton University Press, 2012), 3, 12, 14–15.
13. Amara Miller, "The Origins of Yoga: Part III," *The Sociological Yogi*, September 23, 2014. https://amaramillerblog.wordpress.com/2014/05/29/the-origins-of-yoga-part-iii/.
14. Amara Miller, "The Origins of Yoga: Part III," *The Sociological Yogi*, May 29, 2014, https://amaramillerblog.wordpress.com/2014/05/29/the-origins-of-yoga-part-iii/.
15. Vivekananda, *Karma Yoga*, 81.
16. "The Indigeneity of Trauma Focused Yoga with Lakshmi Nair," *Studio Ānanda*, September 2, 2012. https://www.studioananda.space/2021/09/02/the-indigeneity-of-trauma-focused-yoga-with-lakshmi-nair/.
17. Robin Wall Kimmerer, *Braiding Sweetgrass* (Tantor Media, Inc., 2016).
18. Robin Wall Kimmerer, "The Serviceberry," *Emergence Magazine*, https://emergencemagazine.org/essay/the-serviceberry/.

Chapter 8: On IBS

1. Bethany Webster, *Discovering the Inner Mother: A Guide to Healing the Mother Wound and Claiming Your Personal Power* (William Morrow, 2021), 39.
2. Webster, *Discovering the Inner Mother*, 39.
3. Webster, *Discovering the Inner Mother*, 39.
4. joanna hedva, "Sick Woman Theory." Accessed February 15, 2022. https://johannahedva.com/SickWomanTheory_Hedva_2020.pdf.
5. Gloria Anzaldúa, *Light in the Dark/Luz En Lo Oscuro: Rewriting Identity, Spirituality, Reality* (Duke University Press, 2015), 24.
6. Leah Lakshmi Piepzna-Samarasinha, *Care Work: Dreaming Disability Justice* (Arsenal Pulp Press, 2021), 56.

Chapter 9: Introduction to Radical Self-Care

1. Sarah Maslin Nir, "The Price of Nice Nails," *New York Times*, May 7, 2015. https://www.nytimes.com/2015/05/10/nyregion/at-nail-salons-in-nyc-manicurists-are-underpaid-and-unprotected.html.
2. Brontë Velez and Tricia Hersey, "Resting on and for the Earth," *Atmos*, April 28, 2021.
3. CJ Hauser, "The Crane Wife," *The Paris Review*, August 19, 2019. https://www.theparisreview.org/blog/2019/07/16/the-crane-wife/.
4. joanna hedva, "Sick Woman Theory." Accessed February 15, 2022. https://johannahedva.com/SickWomanTheory_Hedva_2020.pdf.
5. Mia Mingus, "The Four Parts of Accountability: How to Give a Genuine

Apology, Part 1," *Leaving Evidence,* December 18, 2019. https://leavingevidence.wordpress.com/2019/12/18/how-to-give-a-good-apology-part-1-the-four-parts-of-accountability/.
6. Audre Lorde, *The Cancer Journals* (Spinsters/Aunt Lute, 1987), 24.
7. Ann Cvetkovich, "Depression Is Ordinary: Public Feelings and Saidiya Hartman's Lose Your Mother," *Feminist Theory* 13, no. 2 (August 2012): 131–46. https://doi.org/10.1177/1464700112442641.
8. Henri J. M. Nouwen, *Out of Solitude: Three Meditations on the Christian Life* (Ave Maria Press, 2004), 18.
9. Louise A. DeSalvo, *Writing as a Way of Healing: How Telling Our Stories Transforms Our Lives* (Beacon Press, 2000), 185.
10. DeSalvo, *Writing as a Way of Healing,* 185.
11. joanna hedva, "Sick Woman Theory."
12. Gloria Anzaldúa, *Light in the Dark/Luz en Lo Oscuro: Rewriting Identity, Spirituality, Reality* (Duke University Press, 2015), 29.
13. Clarissa Pinkola Estés, *Women Who Run with the Wolves: Myths and Stories of the Wild Woman Archetype* (Ballantine Books, 1995), 84.

Chapter 10: On Self-Care and Self-Harm

1. Clarissa Pinkola Estés, *Women Who Run with the Wolves: Myths and Stories of the Wild Woman Archetype* (Ballantine Books, 1995), 51.
2. Judith Lewis Herman, *Trauma and Recovery: The Aftermath of Violence, from Domestic Abuse to Political Terror* (Basic Books, 2015), 13.
3. Herman, *Trauma and Recovery,* 14.
4. Herman, *Trauma and Recovery,* 14.
5. joanna hedva, "Sick Woman Theory." Accessed February 15, 2022. https://johannahedva.com/SickWomanTheory_Hedva_2020.pdf.
6. Herman, *Trauma and Recovery,* 52.
7. Herman, *Trauma and Recovery,* 86.
8. Herman, *Trauma and Recovery,* 104.
9. Herman, *Trauma and Recovery,* 106.
10. Herman, *Trauma and Recovery,* 109.
11. Herman, *Trauma and Recovery,* 109.
12. Tamara Santibañez, *Could This Be Magic? Tattooing as Liberation Work* (Afterlife Press, 2021), 17.
13. Santibañez, *Could This Be Magic?,* 16.
14. Bethany Webster, *Discovering the Inner Mother: A Guide to Healing the Mother Wound and Claiming Your Personal Power* (William Morrow, 2021), 139.

Chapter 11: On Eroticism

1. Leïla Slimani, *Sex and Lies: True Stories of Women's Intimate Lives in the Arab World,* trans. Sophie Lewis (Penguin Books, 2020), 34.
2. Slimani, *Sex and Lies,* 19.

3. Slimani, *Sex and Lies*, 20.
4. Slimani, *Sex and Lies*, 94.
5. Slimani, *Sex and Lies*, 94.
6. Pankaj Mishra, *From the Ruins of Empire: The Intellectuals Who Remade Asia* (Picador Paper, 2013), 9.
7. William Belsham, *Memoirs of the Reign of George III: From His Accession to the Peace of Amiens*, Vol. 3 (Nabu Press, 2010), 451.
8. Erica Ekrem, "Transcript: Ella Noah Bancroft on the Intelligence of Our Intimacy / 224," *For the Wild*, March 5, 2021. https://forthewild.world/podcast-transcripts/ella-noah-bancroft-on-the-intelligence-of-our-intimacy-224.
9. Swami Satyananda Saraswati, *Kundalini Tantra* (Satyananda Ashram, 1985).
10. Pankaj Mishra, *From the Ruins of Empire*, 101.
11. Shridhar Sharma, "Psychiatry, Colonialism and Indian Civilization: A Historical Appraisal," *Indian Journal of Psychiatry* 48, no. 2 (2006): 109–12. https://doi.org/10.4103/0019-5545.31600.
12. Slimani, *Sex and Lies*, 106.
13. Paulo Freire, *Pedagogy of the Oppressed* (Continuum, 2005), 140.
14. Mishra, *From the Ruins of Empire*, 77.
15. Freire, *Pedagogy of the Oppressed*, 55.
16. Slimani, *Sex and Lies*, 38.
17. Slimani, *Sex and Lies*, 95.
18. Slimani, *Sex and Lies*, 118.
19. Roberto Calasso, *Literature and the Gods* (Vintage International/Knopf Doubleday, 2002), 176.
20. Freire, *Pedagogy of the Oppressed*, 69.

Chapter 12: On Divination

1. Kenneth Johnson, "Buran of Baghdad: An Astrological Woman in the Early Middle Ages," *Geocosmic Journal* (Autumn 2006): 29–33. https://theastrologypodcast.com/articles/Johnson-Buran-Of-Baghdad.pdf.
2. Jessica Dore, *Tarot for Change: Using the Cards for Self-Care, Acceptance, and Growth* (Penguin Books, 2021), 7.
3. Dore, *Tarot for Change*, 47.
4. Dore, *Tarot for Change*, 92.
5. Dore, *Tarot for Change*, 43.
6. Swami Satyananda Saraswati, *Kundalini Tantra* (Satyananda Ashram, 1985), 18.
7. Johnson, "Buran of Baghdad," 29–33.

Chapter 13: Who Is Wellness For?

1. Brandon Baker, "The Language of Climate Change—and the Anthropocene," *Penn Today*, February 5, 2019. https://penntoday.upenn.edu/news/language-climate-change-and-anthropocene (accessed 2/15/22).

2. The Red Nation, *The Red Deal: Indigenous Action to Save Our Earth* (Common Notions, 2021), 19.
3. The Red Nation, *The Red Deal*, 20.
4. Sarah Jaquette Ray, "Climate Anxiety Is an Overwhelmingly White Phenomenon," *Scientific American*, March 21, 2021. https://www.scientificamerican.com/article/the-unbearable-whiteness-of-climate-anxiety/.
5. The Red Nation, *The Red Deal*, 20.
6. Ray, "Climate Anxiety Is an Overwhelmingly White Phenomenon."
7. Robin Wall Kimmerer, *Braiding Sweetgrass* (Tantor Media, Inc., 2016), 307.
8. Ray, "Climate Anxiety Is an Overwhelmingly White Phenomenon."
9. Kimmerer, *Braiding Sweetgrass*, 338.
10. Maxine Bédat, *Unraveled: The Life and Death of a Garment* (New York: Portfolio, 2021), 17.
11. Bédat, *Unraveled*, 19.
12. Maya Tiwari Bri. *The Path of Practice: A Woman's Book of Ayurvedic Healing* (Wellspring/Ballantine, 2001), 72.

Chapter 14: On Degrowth

1. "Gerboise Bleue (Nuclear Test)," Wikipedia, accessed September 8, 2021, https://en.wikipedia.org/wiki/Gerboise_Bleue_(nuclear_test).
2. Malia Bouattia, "France's Nuclear Colonial Legacy in Algeria," *The New Arab*. https://english.alaraby.co.uk/opinion/frances-nuclear-colonial-legacy-algeria.
3. *France's Nuclear Colonial Legacy* by Al Araby.
4. Bouattia, "France's Nuclear Colonial Legacy in Algeria."
5. W.E.B. Dubois, *Darkwater: Voices from within the Veil* (Harcourt, Brace and Howe, 1920), 30.
6. Ben Ehrenreich, "We're Hurtling toward Global Suicide," *The New Republic*, March 18, 2021. https://newrepublic.com/article/161575/climate-change-effects-hurtling-toward-global-suicide; Amitav Ghosh, *The Great Derangement: Climate Change and the Unthinkable* (University of Chicago Press, 2017), 128.
7. Ehrenreich, "We're Hurtling toward Global Suicide."
8. Ehrenreich, "We're Hurtling toward Global Suicide."
9. Anand Giridharadas, *Winners Take All: The Elite Charade of Changing the World* (Alfred A. Knopf, 2018), 18.
10. Silvia Federici, *Enduring Western Civilization: The Construction of the Concept of Western Civilization and Its "Others"* (Praeger, 1995), 76.
11. Agha Shahid Ali, "Farewell," Tumblr, 2012. https://allyourprettywords.tumblr.com/post/28990463502/farewell-agha-shahid-ali.
12. Simar Deol, "Wellness Influencers Owe the Indian Farmers Their Solidarity," Bitch Media, June 1, 2020. https://www.bitchmedia.org/article/wellness-influencers-farmers-protest-solidarity.

13. Arundhati Roy, *The End of Imagination* (Haymarket Books, 2016), 243.
14. The Red Nation, *The Red Deal: Indigenous Action to Save Our Earth* (Common Notions, 2021), 44.
15. The Red Nation, *The Red Deal,* 113.
16. Tyson Yunkaporta, *Sand Talk: How Indigenous Thinking Can Save the World* (HarperOne, 2020), 71.
17. Vandana Shiva, *Earth Democracy: Justice, Sustainability, and Peace* (North Atlantic Books, 2015), 37.
18. Vandana Shiva, "Seed Freedom: Toward an Earth Democracy, A Conversation with Vandana Shiva," Interview by LinYee Yuan, *Mold,* November 11, 2021. https://thisismold.com/mold-magazine/vandana-shiva-seed-freedom-toward-an-earth-democracy.
19. Shiva, "Seed Freedom."
20. Shiva, *Earth Democracy,* 24.
21. The Red Nation, *The Red Deal,* 108.
22. The Red Nation, *The Red Deal,* 99.
23. "American Indian Women and Reproductive Justice." Indigenous Women Rising, March 25, 2017. https://www.iwrising.org/single-post/2017/03/24/american-indian-women-and-reproductive-justice.
24. Maxine Bédat, *Unraveled: The Life and Death of a Garment* (Portfolio, 2021), 36.
25. Bédat, *Unraveled,* 91.
26. Shiva, "Seed Freedom."
27. P. Sainath, "Have India's Farm Suicides Really Declined?," *BBC News,* July 14, 2014. https://www.bbc.com/news/world-asia-india-28205741.
28. Jason Hickel et al., "Rich Countries Drained $152tn from the Global South since 1960." *Al Jazeera,* May 6, 2021. https://www.aljazeera.com/opinions/2021/5/6/rich-countries-drained-152tn-from-the-global-south-since-1960.
29. Hickel et al., "Rich Countries Drained $152tn from the Global South since 1960."
30. Hickel et al., "Rich Countries Drained $152tn from the Global South since 1960."
31. Bédat, *Unraveled,* 43.
32. Bédat, *Unraveled,* 43.
33. Riccardo Mastini, "Degrowth as a Concrete Utopia," *OpenDemocracy,* November 7, 2018. https://www.opendemocracy.net/en/transformation/degrowth-as-concrete-utopia/.
34. Giorgos Kallis, *Degrowth: A Vocabulary for a New Era,* ed. Giacomo D'Alisa and Federico Demaria (Routledge, Taylor & Francis Group, 2015).
35. Bédat, *Unraveled,* 61.
36. Collin Chambers, *Degrowth: An Environmental Ideology with Good Intentions, Bad Politics,* Hampton Institute, July 29, 2021. https://www.hamptonthink.org/read/degrowth-an-environmental-ideology-with-good-intentions-bad-politics.

37. Riccardo Mastini, "Degrowth as a Concrete Utopia."
38. Chambers, *Degrowth*.
39. Mastini, "Degrowth as a Concrete Utopia."
40. Jason Hickel, "How to Achieve Full Decolonization," *New Internationalist*, October 29, 2021. https://newint.org/features/2021/08/09/money-ultimate-decolonizer-fjf.
41. Hickel, "How to Achieve Full Decolonization."
42. Budd L. Hall and Rajesh Tandon, "Decolonization of Knowledge, Epistemicide, Participatory Research and Higher Education," *Research for All* 1, no. 1 (2017): 6–19. https://doi.org/10.18546/RFA.01.1.02.
43. Hall and Tandon, "Decolonization of Knowledge, Epistemicide, Participatory Research and Higher Education.".
44. "Experience Mpambo Afrikan Multiversity," Mpambo Afrikan Multiversity. https://www.mpamboafrikanmultiversity.com/.
45. Emile Durkheim. *Suicide, A Study in Sociology* (Free Press, 1996).
46. The Red Nation, *The Red Deal*, 91.
47. "Tony Morrison on Trauma, Survival, and Finding Meaning," YouTube video, posted by "CTFORUM," November 13, 2020, https://www.youtube.com/watch?v=5xvJYrSsXPA.

Chapter 15: On Healing Our Wounds with the Feminine

1. Gabor Maté. *When the Body Says No: Exploring the Stress-Disease Connection* (Vermilion, 2019), 7.
2. Clarissa Pinkola Estés. *Women Who Run with the Wolves: Myths and Stories of the Wild Woman Archetype* (Ballantine Books, 1995), 188.
3. Bethany Webster, *Discovering the Inner Mother: A Guide to Healing the Mother Wound and Claiming Your Personal Power* (William Morrow, 2021), 250.
4. Webster, *Discovering the Inner Mother*, 250.
5. Sebene Selassie, *You Belong: A Call for Connection* (HarperOne, 2021), 48.
6. The Red Nation. *Red Deal: Indigenous Action to Save Our Earth* (Common Notions, 2021), introduction.
7. William J. Ripple, Christopher Wolf, Thomas M. Newsome, Phoebe Barnard, and William R. Moomaw, "World Scientists' Warning of a Climate Emergency," *BioScience* 70, no 1 (January 2020): 8–12. https://doi.org/10.1093/biosci/biz088.
8. Tyson Yunkaporta. *Sand Talk: How Indigenous Thinking Can Save the World* (HarperOne, 2020), 85.
9. Kamea Chayne, "Bayo Akomolafe Speaks about Slowing Down and Surrendering Human Centrality," *Green Dreamer*, September 16, 2021. https://greendreamer.com/podcast/dr-bayo-akomolafe-the-emergence-network.
10. The Red Nation, *The Red Deal*, 37.
11. The Red Nation, *The Red Deal*, 24.

12. The Red Nation, *The Red Deal*, 87.
13. Yunkaporta, *Sand Talk*.

Conclusion: On Sacred Reciprocity

1. Anna-Sophie Springer and Paulo Tavares, "The Political Nature of the Forest: A Botanic Archaeology of Genocide," *The Word for World Is Still Forest* (K. Verlag, 2017).
2. The Red Nation, *Red Deal: Indigenous Action to Save Our Earth* (Common Notions, 2021), 114.
3. Tyson Yunkaporta, *Sand Talk: How Indigenous Thinking Can Save the World* (HarperOne, 2020), 69.
4. Robin Wall Kimmerer, "The Serviceberry," *Emergence Magazine*. https://emergencemagazine.org/essay/the-serviceberry/.
5. Mia Mingus, "The Four Parts of Accountability: How to Give a Genuine Apology, Part 1," *Leaving Evidence*, July 2, 2020. https://leavingevidence.wordpress.com/2019/12/18/how-to-give-a-good-apology-part-1-the-four-parts-of-accountability/.
6. Kimmerer, "The Serviceberry."
7. Kimmerer, "The Serviceberry."
8. Clarissa Pinkola Estés, *Women Who Run with the Wolves: Myths and Stories of the Wild Woman Archetype* (Ballantine Books, 1995), 201.

BIBLIOGRAPHY

"About." Diaspora Co., https://www.diasporaco.com/pages/about.

"About." The Nap Ministry, March 25, 2021, https://thenapministry.wordpress.com/about/.

Alessandri, Mariana. "The Gender Politics of Fasting." *New York Times*, January 14, 2019, https://www.nytimes.com/2019/01/14/opinion/fasting-gender-politics.html.

Ambedkar, B. R. *Annihilation of Caste*. Rupa, 2018.

"American Indian Women and Reproductive Justice." *Indigenous Women Rising*, March 25, 2017. https://www.iwrising.org/single-post/2017/03/24/american-indian-women-and-reproductive-justic.

Anzaldúa, Gloria. *Light in the Dark/Luz En Lo Oscuro: Rewriting Identity, Spirituality, Reality*. Duke University Press, 2015.

Aziz, Omer. "Blighted by Empire: What the British Did to India." *Los Angeles Review of Books*, 1 Sept. 2018, https://lareviewofbooks.org/article/blighted-by-empire-what-the-british-did-to-india/.

Bald, Vivek. "Unsettling the Nation from the Land: A Conversation with Manu Karuka." Asian American Writers' Workshop, *The Margins*, May 29, 2019. https://aaww.org/unsettling-the-nation-from-the-land-a-conversation-with-manu-karuka/.

Bédat, Maxine. *Unraveled: The Life and Death of a Garment*. Portfolio, 2021,

Belsham, William. *Memoirs of the Reign of George III: From His Accession, to the Peace of Amiens*. Vol. 3. Nabu Press, 2010.

"Biopiracy." Merriam-Webster.com Dictionary, Merriam-Webster, https://www.merriam-webster.com/dictionary/biopiracy.

Bouattia, Malia. "France's Nuclear Colonial Legacy in Algeria." *The New Arab*. https://english.alaraby.co.uk/english/comment/2021/2/12/frances-nuclear-colonial-legacy-in-algeria.

Bouhdiba, Abdelwahab. *Sexuality in Islam*. Saqi, 2004.

Boyer, Anne. "What Is the Language of Pain? An Ugly Gathering of Adjectives." Literary Hub, September 16, 2019. https://lithub.com/anne-boyer-what-is-the-language-of-pain/.

Boulle, Pierre H. "Francois Bernier and the Origins of the Modern Concept of Race." *The Color of Liberty: Histories of Race in France*, edited by Sue Peabody and Tyler Stovell. Duke University Press, 2003, pp. 11–19.

Brennan, Chris, host. "Queen Buran, Astrologer in 9th Century Baghdad." *The Astrology Podcast*. March 12, 2021. https://theastrologypodcast.com/2021/03/12/queen-buran-astrologer-in-9th-century-baghdad/.

Brown, Adrienne Maree. *Pleasure Activism: The Politics of Feeling Good*. AK Press, 2019.

Carpenter, Zoë. "What's Killing America's Black Infants?" *The Nation*, December 23, 2019. https://www.thenation.com/article/archive/whats-killing-americas-black-infants/.

Chambers, Collin. *Degrowth: An Environmental Ideology with Good Intentions, Bad Politics*. Hampton Institute, July 29, 2021. https://www.hamptonthink.org/read/degrowth-an-environmental-ideology-with-good-intentions-bad-politics.

Chayne, Kamea. "Bayo Akomolafe Speaks about Slowing down and Surrendering Human Centrality." *Green Dreamer*, September 16, 2021. https://greendreamer.com/podcast/dr-bayo-akomolafe-the-emergence-network.

"Clean Living Movement." Wikipedia, Wikimedia Foundation. September 25, 2021. https://en.wikipedia.org/wiki Clean_living_movement.

Crenshaw, Kimberlé, host. "What's the Matter with Georgia? Virus, Voting, & Vigilantism the Peach State." *Under the Blacklight*. May 13, 2020.

Cunningham, Vinson. "The Future of L.A. Is Here. Robin D.G. Kelley's Radical Imagination Shows Us the Way." *Los Angeles Times*, March 17, 2021. https://www.latimes.com/lifestyle/image/story/2021–03–17/robin-dg-kelley-black-marxism-protests-la-politics.

Deol, Simar. "Wellness Influencers Owe the Indian Farmers Their Solidarity." Bitch Media, June 1, 2020. https://www.bitchmedia.org/article/wellness-influencers-farmers-protest-solidarity.

DeSalvo, Louise A. *Writing as a Way of Healing: How Telling Our Stories Transforms Our Lives*. Beacon Press, 2000.

Diagnostic and Statistical Manual of Mental Disorders: DSM-5. Arlington, VA: American Psychiatric Association, 2017.

Dominici, Agostino. "The Magic of Thoth-Hermes." New Acropolis Library. September 14, 2018. https://library.acropolis.org/the-magic-of-thoth-hermes/.

Dore, Jessica. *Tarot for Change: Using the Cards for Self-Care, Acceptance, and Growth*. Penguin Books, 2021.

Durkheim, Emile. *Suicide, A Study in Sociology*. New York: Free Press, 1996.

Edgerton, Franklin. *Buddhist Hybrid Sanskrit Grammar and Dictionary*. Yale University Press, 1972.

Ehrenreich, Ben. "We're Hurtling Toward Global Suicide." *The New Republic*. March 18, 2021. https://newrepublic.com/article/161575/climate-change-effects-hurtling-toward-global-suicide.

Eisenstein, Charles. *Sacred Economics: Money, Gift and Society in the Age of Transition*. North Atlantic Books, 2021.

Ekrem, Erica. "Transcript: Ella Noah Bancroft on the Intelligence of Our Intimacy / 224." *For the Wild*, March 5, 2021. https://forthewild.world/podcast-transcripts/ella-noah-bancroft-on-the-intelligence-of-our-intimacy-224.

Ellis, Lisa. "Snapshot of the American Pharmaceutical Industry." Harvard T.H

Chan School of Public Health, October 28, 2019. https://www.hsph.harvard.edu/ecpe/snapshot-of-the-american-pharmaceutical-industry/.

Emoto, Masaru. *The Hidden Messages in Water.* Beyond Words, 2004.

"Enmeshment." Wikipedia. Wikimedia Foundation. January 15, 2022. https://en.wikipedia.org/wiki/Enmeshment.

Estés, Clarissa Pinkola. *Women Who Run with the Wolves: Myths and Stories of the Wild Woman Archetype.* Ballantine Books, 1995.

"Experience Mpambo Afrikan Multiversity." Mpambo Afrikan Multiversity. https://www.mpamboafrikanmultiversity.com/.

Federici, Silvia. *Caliban and the Witch: Women, the Body and Primitive Accumulation.* Autonomedia, 2014.

Federici, Silvia. *Enduring Western Civilization: The Construction of the Concept of Western Civilization and Its "Others."* Praeger, 1995.

"Finding the Cure for Pain in the Pain with Shadi Sankary." Studio Ānanda, https://www.studioananda.space/2021/09/01/finding-the-cure-for-pain-in-the-pain-with-shadi-sankary/.

Forbes, David. "How Capitalism Captured the Mindfulness Industry." *The Guardian,* Guardian News and Media, April 16, 2019. https://www.theguardian.com/lifeandstyle/2019/apr/16/how-capitalism-captured-the-mindfulness-industry.

"France's Nuclear Testing Programme." CTBTO Preparatory Commission. https://www.ctbto.org/nuclear-testing/the-effects-of-nuclear-testing/frances-nuclear-testing-programme/.

Francois Bernier, "Voyage dans les Etats du Grand Mogol", introduction de France Bhattacharya (Arthème Fayard ed. Paris, 1981).

Freire, Paulo. *Pedagogy of the Oppressed.* Continuum, 2005.

"Gabor Maté." Wikipedia. Wikimedia Foundation. December 6, 2021. https://en.wikipedia.org/wiki/Gabor_Mat%C3%A9.

Garner, Dwight. "Inside the List." *New York Times.* March 13, 2005. https://www.nytimes.com/2005/03/13/books/review/inside-the-list.html.

"Gerboise Bleue (Nuclear Test)." Wikipedia. Wikimedia Foundation. September 8, 2021. https://en.wikipedia.org/wiki/Gerboise_Bleue_(nuclear_test).

Giridharadas, Anand. *Winners Take All: The Elite Charade of Changing the World.* Alfred A. Knopf, 2018.

Hall, Budd L., and Rajesh Tandon, "Decolonization of Knowledge, Epistemicide, Participatory Research and Higher Education." *Research for All* 1, no. 1 (2017): 6–19. https://doi.org/10.18546/RFA.01.1.02.

Hartman, Saidiya V. *Scenes of Subjection: Terror, Slavery, and Self-Making in Nineteenth-Century America.* Oxford University Press, 2010.

Hauser, CJ. "The Crane Wife." *The Paris Review,* August 19, 2019. https://www.theparisreview.org/blog/2019/07/16/the-crane-wife/.

Hedva, Johanna. "Sick Woman Theory." Accessed February 15, 2022. https://johannahedva.com/SickWomanTheory_Hedva_2020.pdf.

Herman, Judith Lewis. *Trauma and Recovery: The Aftermath of Violence, From Domestic Abuse to Political Terror*. Basic Books, 2015.

Hickel, Jason. "How to Achieve Full Decolonization." *New Internationalist*, October 29, 2021. https://newint.org/features/2021/08/09/money-ultimate-decolonizer-fjf.

Hickel, Jason, et al. "Rich Countries Drained $152tn from the Global South since 1960." *Al Jazeera*, May 6, 2021. https://www.aljazeera.com/opinions/2021/5/6/rich-countries-drained-152tn-from-the-global-south-since-1960.

Holmes, David Dale. "A Buddhist View on Caste and Equality." Buddhistdoor Global, June 15, 2021. https://www.buddhistdoor.net/features/a-buddhist-view-on-caste-and-equality/.

Illich, Ivan. *Limits to Medicine: Medical Nemesis, the Expropriation of Health*. M. Boyars, 2013.

"The Indigeneity of Trauma Focused Yoga with Lakshmi Nair." Studio Ānanda. https://www.studioananda.space/2021/09/02/the-indigeneity-of-trauma-focused-yoga-with-lakshmi-nair/.

Johnson, Susan M. *Hold Me Tight: Seven Conversations for a Lifetime of Love*. Little, Brown, 2010.

Jordan, June. *Civil Wars*. Simon & Schuster, 1995.

Kallis, Giorgos. *Degrowth: A Vocabulary for a New Era*. Edited by Giacomo D'Alisa and Federico Demaria. Routledge, Taylor & Francis Group, 2015.

Kaushik, R. K. *Architect of Human Destiny?: Who Brings about Peace or Chaos?* Kalpaz Publications, 2003.

Kenneth Johnson, "Buran of Baghdad: An Astrological Woman in the Early Middle Ages." *Geocosmic Journal*. Autumn 2006, pp. 29–33.

Kimmerer, Robin Wall. *Braiding Sweetgrass*. Tantor Media, Inc., 2016.

Kimmerer, Robin Wall. "The Serviceberry: An Economy of Abundance." Global Oneness Project. Accessed February 15, 2022. https://www.globalonenessproject.org/library/essays/serviceberry-economy-abundance.

Kortava, David. "Lost in Thought." *Harper's Magazine*, April 8, 2021. https://harpers.org/archive/2021/04/lost-in-thought-psychological-risks-of-meditation/.

Lorde, Audre. *The Audre Lorde Compendium*. Pandora, 1996.

Lorde, Audre. *The Cancer Journals*. Spinsters/Aunt Lute, 1987.

Macy, Joanna. "Entering the Bardo." *Emergence Magazine*, October 5, 2021. https://emergencemagazine.org/op_ed/entering-the-bardo/.

Mastini, Riccardo. "Degrowth as a Concrete Utopia." OpenDemocracy, November 7, 2018. https://www.opendemocracy.net/en/transformation/degrowth-as-concrete-utopia/.

Maté, Gabor. "Opinion: Time to Stress Cancer Rates Among the Young." *The Globe and Mail*, April 24, 2002. https://www.theglobeandmail.com/opinion/time-to-stress-cancer-rates-among-the-young/article754321/.

Menakem, Resmaa. *My Grandmother's Hands: Racialized Trauma and the Pathway to Mending Our Hearts and Bodies*. Penguin Books, 2021.

Miller, Amara. "The Origins of Yoga: Part III." *The Sociological Yogi*, September 23, 2014. https://amaramillerblog.wordpress.com/2014/05/29/the-origins-of-yoga-part-iii/.

"Mind–Body Dualism." Wikipedia. Wikimedia Foundation. https://en.wikipedia.org/wiki/Mind%E2%80%93body_dualism.

Mingus, Mia. "The Four Parts of Accountability: How to Give a Genuine Apology, Part 1." *Leaving Evidence*, July 2, 2020. https://leavingevidence.wordpress.com/2019/12/18/how-to-give-a-good-apology-part-1-the-four-parts-of-accountability/.

Minh-Ha, Trinh T. *When the Moon Waxes Red: Representation, Gender and Cultural Politics*. Routledge, 2013.

Mishra, Pankaj. *From the Ruins of Empire*. Picador Paper, 2013.

"Missing and Murdered Indigenous Women and Girls." National Indigenous Women's Resource Center. https://www.niwrc.org/mmiwg-awareness.

Naoroji, Dadabhai. *Poverty of India*. Forgotten Books, 2018.

Navlakha, Meera. "For Indians in the Grip of a COVID-19 Crisis, Twitter Is a Desperate Last Resort." *gal-dem*, April 23, 2021. https://gal-dem.com/india-covid-19-twitter/.

Nguyen, Mimi Thi. *The Gift of Freedom: War, Debt, and Other Refugee Passages*. Duke University Press, 2012.

Nir, Sarah Maslin. "The Price of Nice Nails." *New York Times*, May 7, 2015. https://www.nytimes.com/2015/05/10/nyregion/at-nail-salons-in-nyc-manicurists-are-underpaid-and-unprotected.html.

Nouwen, Henri J. M. *Out of Solitude: Three Meditations on the Christian Life*. Ave Maria Press, 2004.

Owens, Lama Rod. *Love and Rage: The Path of Liberation through Anger*. North Atlantic Books, 2020.

Piepzna-Samarasinha, Leah Lakshmi. *Care Work: Dreaming Disability Justice*. Arsenal Pulp Press, 2021.

Puff, Robert. "An Overview of Meditation: Its Origins and Traditions." *Psychology Today*, July 7, 2013. https://www.psychologytoday.com/intl/blog/meditation-modern-life/201307/overview-meditation-its-origins-and-traditions.

Purser, Ronald. "From Neoliberal to Social Mindfulness." Powell's Books. Accessed February 15, 2022. https://www.powells.com/post/original-essays/from-neoliberal-to-social-mindfulness.

Ray, Sarah Jaquette. "Climate Anxiety Is an Overwhelmingly White Phenomenon." *Scientific American*. March 21, 2021. https://www.scientificamerican.com/article/the-unbearable-whiteness-of-climate-anxiety/.

The Red Nation. *The Red Deal: Indigenous Action to Save Our Earth*. Common Notions, 2021.

Reville, William. "The Pseudoscience of Creating Beautiful (or Ugly) Water." *The Irish Times*, February 20, 2013. https://www.irishtimes.com/news/science/the-pseudoscience-of-creating-beautiful-or-ugly-water-1.574583.

Ross, Ashley. "Meditation History: Religious Practice to Mainstream Trend." *Time*, March 9, 2016. https://time.com/4246928/meditation-history-buddhism/.

Roy, Arundhati. *The End of Imagination*. Haymarket Books, 2016.

Sainath, P. "Have India's Farm Suicides Really Declined?" *BBC News*, July 14, 2014. https://www.bbc.com/news/world-asia-india-28205741.

Santibañez, Tamara. *Could This Be Magic? Tattooing as Liberation Work*. Afterlife Press, 2021.

Saraswati, Swami Satyananda. *Kundalini Tantra*. Satyananda Ashram, 1985.

Selassie, Sebene. *You Belong: A Call for Connection*. HarperOne, 2021.

Scarry, Elaine. *The Body in Pain: The Making and Unmaking of the World*. Oxford University Press, 1985.

Sharma, Shridhar. "Psychiatry, Colonialism and Indian Civilization: A Historical Appraisal." *Indian Journal of Psychiatry* 48, no. 2 (2006): 109–12. https://doi.org/10.4103/0019–5545.31600.

Shiva, Vandana. *Earth Democracy: Justice, Sustainability, and Peace*. North Atlantic Books, 2015.

Shiva, Vandana. "Seed Freedom: Toward an Earth Democracy A Conversation with Vandana Shiva." Interview by LinYee Yuan. *Mold*, November 11, 2021. https://thisismold.com/mold-magazine/vandana-shiva-seed-freedom-toward-an-earth-democracy.

Slimani, Leïla. *Sex and Lies: True Stories of Women's Intimate Lives in the Arab World*. Translated by Sophie Lewis. Penguin Books, 2020.

Singh, Julietta. *No Archive Will Restore You*. Punctum Books, 2018.

Singh, Julietta. *Unthinking Mastery: Dehumanism and Decolonial Entanglements*. Duke University Press, 2018.

Slater, Alice. "The US Has Military Bases in 80 Countries. All of Them Must Close." *The Nation*, January 25, 2018. https://www.thenation.com/article/archive/the-us-has-military-bases-in-172-countries-all-of-them-must-close/.

Springer, Anna-Sophie, and Paulo Tavares. "The Political Nature of the Forest: A Botanic Archaeology of Genocide." Essay in *The Word for World Is Still Forest*. K. Verlag, 2017.

Starhawk. *Dreaming the Dark: Magic, Sex, and Politics*. Beacon Press, 1997.

"Storytelling and Survivorship with Amita Swadhin." Studio Ānanda, https://www.studioananda.space/2021/09/01/storytelling-and-survivorship-with-amita-swadhin/.

Strings, Sabrina. *Fearing the Black Body: The Racial Origins of Fat Phobia*. New York University Press, 2019.

Stuurman, Siep. "François Bernier and the Invention of Racial Classification." *History Workshop Journal* 50 (2000): 1–21. http://www.jstor.org/stable/4289688.

Tagore, Rabindranath. *Nationalism*. Penguin Books, 2017.

Thevarajah, Prinita. "How Casteism Manifests in Yoga and Why It's a Problem." *Byrdie*, September 9, 2021. https://www.byrdie.com/casteism-in-yoga-5119378.

Thorpe, Matthew. "12 Benefits of Meditation." *Healthline*, October 27, 2020. https://www.healthline.com/nutrition/12-benefits-of-meditation.

Tiwari, Bri Maya. *The Path of Practice: A Woman's Book of Ayurvedic Healing*. Wellspring/Ballantine, 2001.

Trickett, Penelope K., et al. "The Impact of Sexual Abuse on Female Development: Lessons from a Multigenerational, Longitudinal Research Study." *Development and Psychopathology* 23, no. 2 (2011): 453–76. https://doi.org/10.1017/S0954579411000174.

"Vedas." Wikipedia, Wikimedia Foundation. October 5, 2021. https://en.wikipedia.org/wiki/Vedas.

Velez, Brontë, and Tricia Hersey. "Resting on and for the Earth." *Atmos*, April 28, 2021.

"Victor Bruce, 9th Earl of Elgin." Wikipedia. Wikimedia Foundation. January 26, 2022. https://en.wikipedia.org/wiki/Victor_Bruce,_9th_Earl_of_Elgin.

Vivekananda, Swami. *Karma Yoga: The Yoga of Action*. Vedanta Press—Advaita Ashrama, 2007.

Wang, Jackie. *Carceral Capitalism*. Semiotext(e), 2018.

Webster, Bethany. *Discovering the Inner Mother: A Guide to Healing the Mother Wound and Claiming Your Personal Power*. William Morrow, 2021.

White, Davis Gordon. "Yoga, Brief History of an Idea." Princeton University Press. Accessed February 15, 2022. https://assets.press.princeton.edu/chapters/i9565.pdf.

White, David Gordon, ed. *Yoga in Practice*. Princeton University Press, 2012.

Wischhover, Cheryl. "Can Wellness Be Scientific?" *The Cut*. July 19, 2016. https://www.thecut.com/swellness/2016/07/can-wellness-be-scientific.html.

Wojnarowicz, David. *Close to the Knives: A Memoir of Disintegration*. Canongate Books, 2017.

Yellow Bird, Michael. "Decolonizing the Mind: Using Mindfulness Research and Traditional Indigenous Ceremonies to Delete the Neural Networks of Colonialism." Accessed February 15, 2022. http://www.aihec.org/our-stories/docs/BehavioralHealth/2016/NeurodecolonizationMindfulness_YellowBird.pdf.

Yunkaporta, Tyson. *Sand Talk: How Indigenous Thinking Can Save the World*. HarperOne, 2020.

ABOUT THE AUTHOR

FARIHA RÓISÍN is a multidisciplinary artist, born in Ontario, Canada. She was raised in Sydney, Australia, and is based in Los Angeles, California. As a Muslim queer Bangladeshi, she is interested in the margins, liminality, otherness, and the mercurial nature of being. Her work has pioneered a refreshing and renewed conversation about wellness, contemporary Islam, and queer identities and has been featured in the *New York Times*, Al Jazeera, the *Guardian*, and *Vogue*. She is the author of the poetry collection *How to Cure a Ghost* (2019), as well as the novel *Like a Bird* (2020). Her second book of poetry is entitled *Survival Takes a Wild Imagination*.